Differentiable Manifolds

Differentiable Manifolds

S. T. HU

University of California, Los Angeles

Holt, Rinehart and Winston, Inc.

New York Chicago San Francisco Atlanta Dallas
Montreal Toronto London Sydney

Library of Congress Catalog Card Number: 69–13561
03-075485-2
Printed in the United States of America
1 2 3 4 5 6 7 8 9

DEDICATED TO

S. S. CHERN

Preface

The object of this book is to provide a text for a one-semester course in the subject, essentially following the first fourteen paragraphs of the CUPM recommendations for algebraic topology. The present text can be used either before or after the author's *Homology Theory*, published by Holden-Day, Inc., in 1966, to provide a one-year course in algebraic topology. If the academic year is divided into three quarters, the author's *Cohomology Theory*, published by Markham Publishing Company in 1968, can be used to supplement these two books in the third quarter.

For pedagogical reasons, the text is restricted to finite-dimensional manifolds, although the fundamental materials can be established at no extra cost for manifolds modeled on Banach or Hilbert spaces rather than finite-dimensional Euclidean spaces. Extensions to infinite-dimensional manifolds are left to the reader as exercises. Intelligibility is always preferred to brevity, and unnecessary generalizations are avoided as much as possible.

The bibliography at the end of the book lists a few existing books on the subject together with those books and papers to which references will be made in the text or in the exercises. References to the bibliography are letters, page numbers or statement numbers enclosed in brackets.

Cross references are given in the form (II, §4) or (II, 4.5), where II stands for Chapter II, §4 for Section 4, and 4.5 for the numbering of the statement in the chapter. The chapter numeral will be omitted in case the reference is made in the same chapter.

Certain deviations from standard set-theoretic notations have been adopted in the text; namely, $\square$ is used to denote the empty set and $A\backslash B$ the set-theoretic difference, is usually denoted by $A - B$. The symbol ▌ is used to indicate the end of a proof, and the abbreviation "iff" is used for the phrase "if and only if."

The author is grateful to Professors Victor Klee and Barrett O'Neill for their numerous suggestions, which resulted in significant improvements. It is a further pleasure to acknowledge the financial support from the National Science Foundation while the manuscript was being prepared. Finally, the author wishes to thank the publisher and the printer for their courtesy and cooperation.

Los Angeles, California
January 1969

SZE-TSEN HU

Contents

Differentiable Manifolds

CHAPTER I

Differentiable Manifolds

This introductory chapter is devoted to the definitions of the fundamental concepts about differentiable manifolds, together with their elementary properties.

1. Topological Manifolds

Throughout this text we shall denote by R^n, for any positive integer n, the *n-dimensional Euclidean space*. For the convenience of the reader, we shall briefly recall the definition of R^n as follows:

As a set, R^n consists of all n-tuples $(t_1, \ldots, t_n)$ of real numbers. Hence, R^n is the nth *Cartesian power*

$$R^n = R \times \cdots \times R$$

of the *real line* R. The members of R^n are called *points* of R^n.

For any two points $a = (a_1, \ldots, a_n)$ and $b = (b_1, \ldots, b_n)$ of R^n, let us define their sum $a + b$ by taking

$$a + b = (a_1 + b_1, \ldots, a_n + b_n).$$

This *addition* $+$ makes R^n an *Abelian group* (or *commutative group*). Next define a *scalar multiplication* in R^n by taking

$$\lambda t = (\lambda t_1, \ldots, \lambda t_n)$$

for every real number λ and every point $t = (t_1, \ldots, t_n)$ of R^n. This scalar multiplication makes R^n a *linear space* (or *vector space*) over the field R of real numbers. The *zero element* of this linear space R^n is the *origin* O—that is, the point

$$O = (0, \ldots, 0).$$

Define a function $\nu : R^n \to R$ by taking

$$\nu(t) = \|t\| = \sqrt{\sum_{i=1}^{n} t_i^2}$$

for every point $t = (t_1, \ldots, t_n)$ in R^n. The following conditions of a *norm* in a linear space are satisfied:

(N1) *The triangle inequality:* For any two points a and b in R^n, we have

$$\|a + b\| \leq \|a\| + \|b\|.$$

(N2) For any real number λ and every point t in R^n, we have

$$\|\lambda t\| = |\lambda| \cdot \|t\|.$$

(N3) For any point t in R^n, $\|t\| = 0$ implies $t = 0$.

The condition (N1) is a straightforward consequence of the Schwarz inequality, and the remaining two conditions are obvious from the definition. This norm in R^n is known as the *Euclidean norm*.

Next define a function

$$d : R^n \times R^n \to R$$

by taking

$$d(a,b) = \|a - b\|$$

for every pair (a,b) of points a and b in R^n. This function d satisfies the conditions of a *metric* defined in general topology and is known as the *Euclidean metric* in R^n. This metric d in R^n makes R^n a metric space known as the *n-dimensional Euclidean space* or the *Euclidean n-space.* Its topology defined by d coincides with the *product topology* in R^n.

Now let us define the notion of a *topological manifold.* For this purpose, let n denote any given positive integer. By an *n-dimensional topological manifold,* we mean a nonempty paracompact Hausdorff space X such that every point x of X has an open neighborhood U in X that is homeomorphic to an open subspace of the n-dimensional Euclidean space R^n. Such a neighborhood U of x in X is called a *coordinate neighborhood.* Here, a topological space X is said to be *paracompact* iff every open cover of X has a locally finite refinement. In the sequel, n-dimensional topological manifolds will be simply called *n-manifold.*

Proposition 1.1. *A nonempty paracompact Hausdorff space X is an n-manifold iff every point x of X has an open neighborhood V in X that is homeomorphic to the n-dimensional Euclidean space R^n.*

PROOF. Since R^n is an open subspace of itself, the sufficiency of (1.1) is obvious. It remains to prove its necessity.

For this purpose, let x denote any point of an n-manifold X. By definition, there exists an open neighborhood U of x in X together with an imbedding

$$i : U \to R^n$$

such that $i(U)$ is an open subspace of R^n. Let $y = i(x) \in R^n$. Then $i(U)$ is an open neighborhood of y in R^n. Since the topology of R^n is defined by the Euclidean metric in R^n, there exists a positive real number δ such that the open δ-ball

$$B_\delta(y) = \{t \in R^n \mid \|t - y\| < \delta\}$$

about y is contained in $i(U)$. Let

$$V = i^{-1}[B_\delta(y)].$$

Then V is an open neighborhood of x in X that is homeomorphic to R^n.∎

Because of (1.1), topological manifolds are frequently called *locally Euclidean spaces.*

By the *unit n-ball*, we mean the following subspace of R^n:

$$B^n = \{t \in R^n \mid \|t\| \leq 1\}.$$

Obviously, B^n is compact, convex, and contractible.

Proposition 1.2. *Every neighborhood N of any point x of an n-manifold X contains a neighborhood M of x in X that is homeomorphic to the unit n-ball B^n.*

PROOF. Since X is an n-manifold, there exists an open neighborhood U of x in X together with an imbedding

$$i : U \to R^n$$

such that $i(U)$ is an open subspace of R^n. Let $y = i(x)$ and

$$W = U \cap \text{Int}(N).$$

Then $i(W)$ is an open neighborhood of y in R^n. There exists a positive real number δ such that the closed δ-ball

$$C_\delta(y) = \{t \in R^n \mid \|t - y\| \leq \delta\}$$

about y is contained in $i(W)$. Let

$$M = i^{-1}[C_\delta(y)].$$

Then M is a neighborhood of x in X that is homeomorphic to B^n.∎

Corollary 1.3. *Every n-manifold is locally compact and locally contractible.*

Examples of topological manifolds:

(a) The n-dimensional Euclidean space R^n and the n-dimensional sphere S^n are n-manifolds.

(b) Every nonempty open subspace of an n-manifold is an n-manifold.

(c) The topological sum of disjoint n-manifolds is an n-manifold.

(d) The topological product of an m-manifold and an n-manifold is an $(m + n)$-manifold.

EXERCISES

1A. Introduce a relation $\sim$ in the subspace

$$W = R^{n+1} \setminus \{0\}$$

of the Euclidean $(n + 1)$-space R^{n+1} as follows: Let

$$x = (x_0, x_1, \ldots, x_n) \qquad y = (y_0, y_1, \ldots, y_n)$$

denote any two points of W. Then

$$x \sim y$$

iff there exists a nonzero real number λ such that

$$y_i = \lambda x_i$$

holds for every $i = 0, 1, \ldots, n$. Verify that $\sim$ is an equivalent relation in W. The quotient space

$$P^n(R) = W/\sim$$

of W over this equivalence relation $\sim$ is known as the *n-dimensional real projective space*. Prove that $P^n(R)$ is an n-manifold.

1B. Consider the points of the Euclidean $(2n + 2)$-space R^{2n+2} as the $(n + 1)$-tuples $z = (z_0, z_1, \ldots, z_n)$ of complex numbers. Introduce a relation $\sim$ in the subspace

$$W = R^{2n+2} \setminus \{0\}$$

of R^{2n+2} as follows. Let

$$u = (u_0, u_1, \ldots, u_n) \qquad v = (v_0, v_1, \ldots, v_n)$$

denote any two points of W. Then

$$u \sim v$$

iff there exists a nonzero complex number λ such that

$$v_i = \lambda u_i$$

holds for every $i = 0, 1, \ldots, n$. Verify that $\sim$ is an equivalence relation in W. The quotient space

$$P^n(C) = W/\sim$$

of W over this equivalence relation $\sim$ is known as the *complex projective n-space*. Prove that $P^n(C)$ is a $2n$-manifold.

1C. Define the *quaternion projective space* $P^n(Q)$ by replacing the system C of all complex numbers in the definition of $P^n(C)$ by the system Q of all quaternions. Prove that $P^n(Q)$ is a $4n$-manifold.

1D. Consider the points of the Euclidean n^2-space R^{n^2} as the square matrices of real numbers with n rows and n columns. Prove that the subspace $\mathcal{O}_n$ of R^{n^2} that consists of all orthogonal matrices is a topological manifold. Determine the dimension of $\mathcal{O}_n$. This topological manifold $\mathcal{O}$ is known as the *orthogonal group* of R^n.

1E. By a *Banach space*, we mean a complete normed linear space. Replacing the Euclidean n-space R^n in the definition of n-manifolds by an arbitrary Banach space B, define topological manifolds of infinite dimension. Check whether (1.1)–(1.3) are still valid.

1F. In the definition of a topological manifold X, can we replace paracompactness of Hausdorff's second axiom of countability?

2. Differentiable Structures

Throughout the present section, let X denote an arbitrary n-manifold. By a *chart* on X, we mean an imbedding

$$\phi : U \to R^n$$

of a nonempty open subspace U of X into the Euclidean n-space such that $\phi(U)$ is an open subspace of R^n. The open subspace U of X will be referred to as the *domain* of the chart ϕ, in symbols,

$$U = \text{Dom }(\phi)$$

For every integer $i = 1, 2, \ldots, n$, let

$$p_i : R^n \to R$$

denote the *natural projection*, defined by

$$p_i(t_1, \ldots, t_n) = t_i$$

for every point $(t_1, \ldots, t_n)$ of R^n. Then, for every chart $\phi : U \to R^n$ on X, the composition

$$\phi_i = p_i \circ \phi : U \to R$$

of ϕ and p_i is called the ith *coordinate function* in U with respect to the chart ϕ. For every point x of U, the real number

$$t_i = \phi_i(x)$$

is called the ith *coordinate* of the point x with respect to the chart ϕ. Because of this, the chart $\phi : U \to R^n$ is usually called a *local coordinate system* in U and, for every point $x \in U$, the n real numbers

$$(t_1, \ldots, t_n) = [\phi_1(x), \ldots, \phi_n(x)] = \phi(x) \in R^n$$

are said to be the *coordinates* of the point x with respect to ϕ.

According to the definition of n-manifolds, every point x has an open neighborhood U with a chart $\phi : U \to R^n$. In case the point x is contained in the intersection $U \cap V$ for two charts $\phi : U \to R^n$ and $\psi : V \to R^n$, it is important to know the relations between the two sets of coordinates $\phi(x)$ and $\psi(x)$.

In order to study these relations, we shall first define *differentiability* of various classes C^k for functions defined on an open subspace of R^n. For this purpose, let

$$f : W \to R$$

denote a real-valued function defined on a nonempty open subspace W of R^n.

The function $f : W \to R$ is said to be *of class* C^0 iff it is continuous. For any positive integer k, $f : W \to R$ is said to be *of class* C^k iff f has continuous partial derivatives of all orders $r \leq k$. The function $f : W \to R$ is said to be *of class* C^∞ iff it is of class C^k for every positive integer k. The function $f : W \to R$ is said to be *of class* C^ω iff it is an analytic function.

Throughout the remainder of the present section, let k denote any member of the set

$$K = \{0, 1, \ldots, \infty, \omega\}.$$

A function

$$f : W \to R^n$$

defined on a nonempty open subspace W of R^n into R^n itself is said to be *of class* C^k iff, for every integer $i = 1, 2, \ldots, n$, the composed function

$$f_i = p_i \circ f : W \to R$$

of f and the natural projection $p_i : R^n \to R$ is of class C^k.

Now let us return to the n-manifold X.

By an *atlas of class* C^k on the n-manifold X, we mean a collection α of charts on X satisfying the following two conditions:

(A1) The domains of the charts in α cover the n-manifold X; in symbols,

$$\bigcup_{\phi \in \alpha} \text{Dom}(\phi) = X.$$

(A2) For any two charts $\phi : U \to R^n$ and $\psi : V \to R^n$ in α with $U \cap V \neq \Box$, the function

$$f_{(\phi,\psi)} : \phi(U \cap V) \to R^n$$

defined by

$$f_{(\phi,\psi)}(t) = \psi[\phi^{-1}(t)]$$

for every point t of $\phi(U \cap V)$ is of class C^k.

The function $f_{(\phi,\psi)}$ in (A2) will be referred to as the *connecting function* of the two charts ϕ and ψ. For every point $x \in U \cap V$, we have

$$f_{(\phi,\psi)}[\phi(x)] = \psi(x).$$

Because of this, $f_{(\phi,\psi)}$ is usually called the *transformation for the change of local coordinate system from* ϕ *to* ψ.

Since $f_{(\phi,\psi)}$ is always continuous, we have the following proposition.

Proposition 2.1. *Every n-manifold has an atlas of class* C^0.

Now let us denote by the symbol

$$\mathfrak{A}^k(X)$$

the set of all atlases on X of class C^k. If $k \neq 0$, this set $\mathfrak{A}^k(X)$ may be empty.

Define a relation $\sim$ in the set $\mathfrak{A}^k(X)$ as follows: For any two atlases α and β in $\mathfrak{A}^k(X)$, $\alpha \sim \beta$ iff $\alpha \cup \beta$ is an atlas in $\mathfrak{A}^k(X)$. In other words, $\alpha \sim \beta$ iff the condition (A2) is satisfied for any two charts $\phi : U \to R^n$ and $\psi: V \to R^n$ in the union $\alpha \cup \beta$ with $U \cap V \neq \square$.

This relation $\sim$ in $\mathfrak{A}^k(X)$ is obviously reflexive, symmetric, and transitive. Hence $\sim$ is an equivalence relation in $\mathfrak{A}^k(X)$ and divides the set $\mathfrak{A}^k(X)$ into disjoint equivalence classes. Each of these equivalence classes is called a *differentiable structure of class* C^k on the given n-manifold X.

The following proposition is obvious.

Proposition 2.2. *Any two atlases of class* C^0 *on an n-manifold are equivalent.*

The following corollary is a direct consequence of (2.1) and (2.2).

Corollary 2.3. *There exists a unique structure of class* C^0 *on any given n-manifold.*

In case $k \neq 0$, an n-manifold X may have no differentiable structure of class C^k or many differentiable structures of class C^k; see [Mi].

Now let us consider an arbitrary differentiable structure σ of class C^k on the given n-manifold X. According to the definition, σ is an equivalence class of atlases in $\mathfrak{A}^k(X)$. Let μ_σ denote the union of all atlases in σ; in symbols,

$$\mu_\sigma = \bigcup_{\alpha \in \sigma} \alpha.$$

Lemma 2.4. *The collection* μ_σ *is an atlas in* $\mathfrak{A}^k(X)$.

PROOF. Since (A1) is clearly satisfied by μ_σ, it remains for (A2) to be verified. For this purpose, consider any two charts

$$\phi : U \to R^n \qquad \psi : V \to R^n$$

in the collection μ_σ. By definition of μ_σ, there are $\alpha \in \sigma$ and $\beta \in \sigma$ with

$$\phi \in \alpha \qquad \psi \in \beta.$$

Since $\alpha \sim \beta$, it follows that (A2) holds.∎

Lemma 2.5. *For every $\alpha \in \sigma$, we have $\mu_\sigma \sim \alpha$.*

PROOF. Since $\mu_\sigma \cup \alpha = \mu_\sigma$, it follows from the definition of $\sim$ that we have $\mu_\sigma \sim \alpha$. ∎

Consequently, μ_σ is an atlas in the differentiable structure σ. Since μ_σ contains every atlas α in σ, it will be referred to as the *maximum atlas* in the differentiable structure σ. Because of the uniqueness of μ_σ, we may even identify the differentiable structure σ with its unique maximum atlas μ_σ.

By a *differentiable n-manifold of class C^k*, we mean an n-manifold X together with a given differentiable structure σ of class C^k on X.

For the special case $k = 0$, it follows from (2.3) that this notion reduces to the notion of an n-manifold. On the other hand, differentiable n-manifolds of class C^ω are usually called *(real) analytic n-manifolds.*

Examples of differentiable manifolds:

(a) Consider the Euclidean n-space R^n. The identity map $i : R^n \to R^n$ is a chart on the n-manifold R^n and the collection $\alpha = \{i\}$ is an atlas of class C^k for every $k \in K$. For every $k \in K$, this atlas $\alpha = \{i\}$ on R^n determines a differentiable structure σ of class C^k on R^n. The maximum atlas μ_σ in σ consists of all charts $\phi : U \to R^n$ on R^n that are of class C^k.

(b) Consider the unit n-sphere S^n in the Euclidean $(n + 1)$-space R^{n+1}. For every point $x \in S^n$, the *stereographic projection*

$$\phi_x : S^n \setminus x \to R^n$$

from the open set $S^n \setminus x$ of S^n onto R^n is a chart on S^n. The collection

$$\alpha = \{\phi_x \mid x \in S^n\}$$

is an atlas of class C^k for every $k \in K$. For every $k \in K$, this atlas α on S^n determines a differentiable structure σ of class C^k on S^n.

(c) Every nonempty open subspace of a differentiable n-manifold of class C^k is a differentiable n-manifold of class C^k.

(d) The topological sum of disjoint differentiable n-manifolds of class C^k is a differentiable n-manifold of class C^k.

(e) The topological product of a differentiable m-manifold of class C^k and a differentiable n-manifold of class C^k is a differentiable $(m + n)$-manifold of class C^k.

In Examples (**c**)–(**e**), the differentiable structures are defined in the obvious ways.

With the understanding that $\infty < \omega$, the following proposition is obvious.

Proposition 2.6. *Every differentiable atlas of class C^k on an n-manifold X is also a differentiable atlas of C^h for every $h \leq k$.*

Because of (2.6), every differentiable n-manifold X of class C^k can be considered as a differentiable n-manifold of class C^h for every $h \leq k$.

Hereafter, differentiable n-manifolds of class C^∞ will be simply called *smooth n-manifolds.*

EXERCISES

2A. Construct atlases on the real projective space $P^n(R)$, the complex projective space $P^n(C)$, and the quaternion projective space $P^n(Q)$ so that they become differentiable manifolds of class C^k for every k.

2B. Prove that the orthogonal group $\mathfrak{O}$ of R^n defined in Exercise 1D can be made a differentiable manifold of class C^k for every k.

2C. Prove that the subspace S of the Euclidean space R^3, which consists of all points (x,y,z) of R^3 satisfying

$$x^3 + y^3 + z^3 - 3xyz = 1$$

can be made a differentiable 2-manifold of class C^k for every k.

2D. Prove that the subspace C of the Euclidean space R^3, which consists of all points (x,y,z) of R^3 satisfying

$$\begin{aligned} x^2 - y^2 + 2xz - 2yz &= 0 \\ 2x - y + z &= 0 \end{aligned}$$

can be made a differentiable 1-manifold of class C^k for every k.

2E. Generalize the notion of differentiation to functions $\phi : U \to Y$ from an open subspace U of a Banach space X into a Banach space Y. Define differentiability of class C^k by induction on k. Then define the notion of infinite-dimensional differentiable manifolds of class C^k, with $k = 0, 1, \ldots, \infty$.

3. Differentiable Functions

Let X denote a differentiable m-manifold of class C^k with differentiable structure ξ. Let Y denote a differentiable n-manifold of class C^k with differentiable structure η, where k is any member of the set

$$K = \{0, 1, \ldots, \infty, \omega\}$$

with the understanding that $\infty < \omega$. Consider an arbitrary function

$$f : X \to Y.$$

Let h denote any member of the set K satisfying $h \leq k$.

The function $f : X \to Y$ is said to be *of class* C^h iff, for every chart $\phi : U \to R^m$ in the maximum atlas μ_ξ of ξ and every chart $\psi : V \to R^n$ in the maximal atlas μ_η of η with

$$W = U \cap f^{-1}(V) \neq \square$$

the function

$$f_{(\phi,\psi)} : \phi(W) \to R^n$$

defined by

$$f_{(\phi,\psi)}(t) = \psi\{f[\phi^{-1}(t)]\}$$

for every point t of the open subspace $\phi(W)$ of the Euclidean m-space R^m, is of class C^h. We recall that $f_{(\phi,\psi)}$ is of class C^h iff the composition

$$p_i \circ f_{(\phi,\psi)} : \phi(W) \to R$$

of $f_{(\phi,\psi)}$ and the natural projection $p_i : R^n \to R$ is of class C^h for every $i = 1, 2, \ldots, n$.

Proposition 3.1. *The function* $f: X \to Y$ *is of class* C^h *iff* ξ *contains an atlas* α *on* X *and* η *contains an atlas* β *on* Y *such that, for every chart* $\phi : U \to R^m$ *in* α *and every chart* $\psi : V \to R^n$ *in* β *with*

$$W = U \cap f^{-1}(V) \neq \square$$

the function $f_{(\phi,\psi)} : \phi(W) \to R^n$ *is of class* C^h.

PROOF. Since $\alpha \subset \mu_\xi$ and $\beta \subset \mu_\eta$, the condition in (3.1) is obviously necessary. It remains to establish its sufficiency.

For this purpose, let $\rho : E \to R^m$ denote any chart in μ_ξ and $\theta : F \to R^n$ denote any chart in μ_η with

$$G = E \cap f^{-1}(F) \neq \square$$

It suffices to prove that the function $f_{(\rho,\theta)} : \rho(G) \to R^n$ is of class C^h.

Since α and β are atlases, we have

$$E = \bigcup_{\phi \in \alpha} [E \cap \text{Dom}(\phi)],$$
$$F = \bigcup_{\psi \in \beta} [F \cap \text{Dom}(\psi)].$$

Consequently, it suffices to prove that, for every chart $\phi : U \to R^m$ in α and every chart $\psi : V \to R^n$ in β with

$$W = U \cap f^{-1}(V) \neq \square,$$

the restriction $f_{(\rho,\theta)} \mid \rho(G \cap W)$ of the function $f_{(\rho,\theta)}$ to the open subspace

$$\rho(G \cap W) = \rho[(E \cap U) \cap f^{-1}(F \cap V)]$$

of the Euclidean m-space R^m is of class C^h.

Since both ρ and ϕ are in μ_ξ, the function

$$f_{(\rho,\phi)} : \rho(E \cap U) \to R^m$$

defined by

$$f_{(\rho,\phi)}(t) = \phi[\rho^{-1}(t)]$$

for every point t of $\rho(E \cap U)$ is of class C^k. Similarly, since both ψ and θ are in μ_η, the function

$$f_{(\psi,\theta)} : \psi(F \cap V) \to R^n$$

defined by

$$f_{(\psi,\theta)}(t) = \theta[\psi^{-1}(t)]$$

for every point t of $\psi(F \cap V)$ is of class C^k.

On the other hand, since $\phi \in \alpha$ and $\psi \in \beta$, it follows from the condition in (3.1) that the function

$$f_{(\phi,\psi)} : \phi(W) \to R^n$$

is of class C^h.

Since $h \leq k$ and

$$f_{(\rho,\theta)}(t) = \theta\{f[\rho^{-1}(t)]\} = f_{(\psi,\theta)}\{f_{(\phi,\psi)}[f_{(\rho,\phi)}(t)]\}$$

for every point t of $\rho(G \cap W)$, it follows that the restriction $f_{(\rho,\theta)} \mid \rho(G \cap W)$ is of class C^h. This completes the proof of (3.1).||

For every $h \leq k$, let us use the symbol

$$\text{Fun}^h(X,Y)$$

to denote the set of all functions $f : X \to Y$ of class C^h. In particular,

$$\text{Fun}^0(X,Y)$$

consists of all continuous functions $f : X \to Y$, usually called *maps* or *mappings*.

For the special case where Y is the real line R, we shall use the following simpler notation:

$$\text{Fun}^h(X) = \text{Fun}^h(X,R).$$

In this case, $\text{Fun}^h(X)$ is an *algebra* over the ring R of real numbers. In fact, for arbitrary real numbers $a, b \in R$ and arbitrary functions $f, g \in \text{Fun}^h(X)$, the functions

$$af + bg : X \to R \qquad fg : X \to R$$

defined by

$$(af + bg)(x) = a[f(x)] + b[g(x)]$$
$$(fg)(x) = f(x)g(x)$$

for every $x \in X$, are also in $\text{Fun}^h(X)$.

Now let us generalize the preceding notion of *differentiability of class* C^h to functions

$$f : A \to Y$$

defined on a nonempty subspace A of X.

If A is an open subspace of X, then A is a differentiable m-manifold of class C^k with the differentiable structure $\xi \mid A$ induced by ξ in the obvious way. In this case, the function $f : A \to Y$ is said to be *of class* C^h iff it is of class C^h with respect to the differentiable structures $\xi \mid A$ in A and η in Y.

For the general case, the function $f : A \to Y$ is said to be *of class* C^h iff it can be extended to a function of class C^h over some open neighborhood of A in X. In other words, the function $f : A \to Y$ is of class C^h iff there exists an open neighborhood U of A in X together with a function

$$g : U \to Y$$

of class C^h such that $g(x) = f(x)$ holds for every $x \in A$.

In particular, X may be the Euclidean m-space R^m. Hence we have also defined the *differentiability of class* C^h for the functions defined on a subspace of R^m into a differentiable manifold of class C^k with $k \geq h$.

The following proposition is obvious.

Proposition 3.2. *If $f : A \to Y$ is a function of class C^h, then f is of class C^g for every $g \leq h$.*

In particular, we have the following corollary of (3.2).

Corollary 3.3. *Every function $f : A \to Y$ of class C^h is continuous.*

The following two propositions are also obvious.

Proposition 3.4. *The identity function $i : X \to X$ on a differentiable n-manifold X of class C^k is of class C^k.*

Proposition 3.5. *The composition*

$$g \circ f : X \to Z$$

of any two functions $f : X \to Y$ and $g : Y \to Z$ of class C^h is of class C^h.

Next let us define the important concept of diffeomorphisms.

Let X and Y denote differentiable n-manifolds of class C^k, and let $h \leq k$. By a *diffeomorphism* of class C^h from X onto Y, we mean a homeomorphism

$$f : X \to Y$$

such that both f and its inverse $f^{-1} : Y \to X$ are functions of class C^h.

The following three propositions are obvious.

Proposition 3.6. *The identity function $i : X \to X$ on a differentiable n-manifold X of class C^k is a diffeomorphism of class C^k.*

Proposition 3.7. *The composition*

$$g \circ f : X \to Z$$

of any two diffeomorphisms $f : X \to Y$ and $g : Y \to Z$ of class C^h is a diffeomorphism of class C^h.

Proposition 3.8. *If $f : X \to Y$ is a diffeomorphism of class C^h, then so is its inverse $f^{-1} : Y \to X$.*

Hereafter, functions of class C^∞ will be called *smooth functions,* and diffeomorphisms of class C^∞ will be simply called *diffeomorphisms.*

The remainder of the section is devoted to the existence of partitions of unity.

Let $\mathfrak{C}$ denote any given open cover of a differentiable n-manifold of class C^k. By a *partition of unity of class C^h, $h \leq k$, subordinate to* $\mathfrak{C}$, we mean a locally finite open refinement $\mathfrak{D}$ of $\mathfrak{C}$ and an indexed family

$$\{f_V : X \to R \mid V \in \mathfrak{D}\}$$

of functions of class C^h satisfying the conditions:

(PU1) For every $V \in \mathfrak{D}$ and every $x \in X$, we have $f_V(x) \geq 0$.

(PU2) For every $V \in \mathfrak{D}$, the *support*

$$S(f_V) = \mathrm{Cl}\{x \in X \mid f_V(x) \neq 0\}$$

of f_V is contained in V.

(PU3) For every $x \in X$, we have

$$\sum_{V \in \mathfrak{D}} f_V(x) = 1.$$

Partitions of unity of class C^0 are simply called *partitions of unity,* and those of class C^∞ will be referred to as *smooth partitions of unity.*

Lemma 3.9. *For any two real numbers a and b satisfying $0 \leq a < b$, there exists a smooth function $f : R^n \to R$ such that*

$$f(x) = \begin{cases} 1 & (\textit{if } \|x\| \leq a) \\ 0 & (\textit{if } \|x\| \geq b) \end{cases}$$

and that $0 \leq f(x) \leq 1$ holds for every $x \in R^n$.

PROOF. Let $\alpha = a^2$ and $\beta = b^2$. Consider the smooth function $\phi : R \to R$ defined by

$$\phi(t) = \begin{cases} \exp\left(\dfrac{1}{t-\beta} - \dfrac{1}{t-\alpha}\right) & (\text{if } \alpha < t < \beta) \\ 0 & (\text{otherwise}). \end{cases}$$

Next define a smooth function $\Phi : R \to R$ by taking

$$\Phi(x) = \int_x^\beta \phi(t)\,dt \Big/ \int_\alpha^\beta \phi(t)\,dt$$

for every $x \in R$. Then Φ has the property that

$$\Phi(x) = \begin{cases} 1 & (\text{if } x \leq a^2) \\ 0 & (\text{if } x \geq b^2) \end{cases}$$

and that $0 \leq \Phi(x) \leq 1$ holds for every $x \in R$. Consequently, the smooth function $f : R^n \to R$ defined by

$$f(x) = \Phi(x_1^2 + \cdots + x_n^2)$$

for every $x = (x_1, \ldots, x_n)$ has the properties in (3.9).▮

Lemma 3.10. *If V is a neighborhood of a compact subset A of a differentiable n-manifold X of class C^k, then for every $h \leq \min\{k, \infty\}$ there exists a function $f : X \to R$ of class C^h satisfying the conditions:*

(a) *For every $x \in X$, we have $0 \leq f(x) \leq 1$.*
(b) *For every $x \in A$, we have $f(x) = 1$.*
(c) *The support $S(f) = \mathrm{Cl}\{x \in X \mid f(x) \neq 0\}$ of f is contained in V.*

PROOF. Let p denote an arbitrary point of A. Select a chart $\phi_p : U_p \to R^n$ in the maximum atlas μ_ξ of the differentiable structure ξ of X. We may assume that $\phi_p(p) = 0$. Since the open set $\phi_p(V \cap U_p)$ of R^n contains the origin $\phi_p(p) = 0$, there exists a positive real number b_p such that, for every $x \in R^n$, $\|x\| \leq b_p$ implies $x \in \phi_p(V \cap U_p)$. Let $a_p = b_p/2$. Then we have

$$0 < a_p < b_p.$$

According to (3.9), there exists a function $f_p : R^n \to R$ of class C^h such that

$$f_p(x) = \begin{cases} 1 & (\text{if } \|x\| \leq a_p) \\ 0 & (\text{if } \|x\| \geq b_p) \end{cases}$$

and that $0 \leq f_p(x) \leq 1$ holds for every $x \in R^n$. Define a function $g_p : X \to R$ by taking

$$g_p(x) = \begin{cases} f_p[\phi_p(x)] & (\text{if } x \in U_p), \\ 0 & (\text{if } x \in X \setminus U_p). \end{cases}$$

Then g_p is clearly of class C^h. Do these for all $p \in A$.

For every $p \in A$, let W_p denote the open neighborhood of p in X defined by

$$W_p = \{x \in U_p \mid \|\phi_p(x)\| < a_p\}.$$

Since A is compact, there exists a finite subset $F = \{p_1, \ldots, p_m\}$ of A such that

$$A \subset W_{p_1} \cup \cdots \cup W_{p_m}.$$

Define a function $f : X \to R$ by taking

$$f(x) = 1 - [1 - g_{p_1}(x)] \cdots [1 - g_{p_m}(x)]$$

for every $x \in X$. Then f is of class C^h and has the properties stated in (3.10).∎

Theorem 3.11. *For any given open cover* $\mathcal{C}$ *of a differentiable n-manifold* X *of class* C^k *and every* $h \leq \min(k, \infty)$, *there exists a partition of unity of class* C^h *that is subordinate to* $\mathcal{C}$.

PROOF. Let x denote any point in X. Since $\mathcal{C}$ is an open cover of X, there is a $U_x \in \mathcal{C}$ with $x \in U_x$. Since X is locally compact, U_x contains a compact neighborhood K_x of x. Do this for all $x \in X$. Since X is paracompact, the open cover

$$\mathfrak{F} = \{\text{Int}\,(K_x) \mid x \in X\}$$

of X has a locally finite open refinement $\mathfrak{D}$. Since $\mathfrak{F}$ is a refinement of $\mathcal{C}$, so is $\mathfrak{D}$.

Since X is normal and $\mathfrak{D}$ is locally finite, we can choose for each $V \in \mathfrak{D}$ an open subset W_V of V such that

$$\text{Cl}(W_V) \subset V$$

and that the collection $\{W_V \mid V \in \mathfrak{D}\}$ covers X. Since $\text{Cl}(W_V)$ is a closed subset of some compact set K_x, $\text{Cl}(W_V)$ is compact. It follows from (3.10) that there exists a function $g_V : X \to R$ of class C^h such that

$$\begin{aligned} 0 \leq g_V(x) &\leq 1 \qquad (x \in X) \\ g_V(x) &= 1 \qquad (x \in A) \\ S(g_V) &\subset V. \end{aligned}$$

Since $\mathfrak{D}$ is locally finite, we may define a function $g : X \to R$ of class C^h by taking

$$g(x) = \sum_{V \in \mathfrak{D}} g_V(x)$$

for every $x \in X$. Since $\{W_V \mid V \in \mathfrak{D}\}$ covers X, we have $g(x) \geq 1$ for every

$x \in X$. Hence we may define for every $V \in \mathfrak{D}$ a function

$$f_V : X \to R$$

by taking

$$f_V(x) = g_V(x)/g(x)$$

for every $x \in X$. The conditions (PU1)–(PU3) are clearly satisfied. This completes the proof of (3.11).▌

To conclude the present section, we note that the statements for the class C^ω corresponding to (3.9)–(3.11) are false. In fact, an analytic function $f : R^n \to R$ must be identically zero on R^n if $f(x) = 0$ holds for every point x of a nonempty open set U of R^n.

EXERCISES

3A. Let X denote any differentiable n-manifold of class C^k with ξ as its differentiable structure. Prove that every chart $\phi : U \to R^n$ in the maximum atlas μ_ξ of ξ is of class C^k if R^n is considered as a differentiable n-manifold as in Example (a) of §2. Hence the ith coordinate function $\phi_i : U \to R$ is of class C^k for every $i = 1, 2, \ldots, n$.

3B. Prove that a map $f : X \to Y$ is of class C^h iff, for every function $\lambda : V \to R$ of class C^h on any nonempty open subspace V of Y, the function $\kappa : U \to R$ defined on the open subspace $U = f^{-1}(V)$ of X by $\kappa(x) = \lambda[f(x)]$ for every $x \in U$ is of class C^h.

3C. Prove that any two differentiable structures ξ and η of class C^k on an n-manifold X are the same iff the identity function $i : X \to X$ is a diffeomorphism of class C^k from (X,ξ) onto (X,η).

3D. Prove that the function $f : R^2 \to R^2$ defined by

$$f(x,y) = (xe^y + y, xe^y - y)$$

for every point (x,y) of R^2 is a diffeomorphism.

3E. Generalize the concepts in the text of this section to the case where X and Y may be infinite-dimensional.

4. Tangent and Cotangent Spaces

Throughout the remainder of the book, we shall confine our study to the class C^∞. This is done mainly for the sake of convenience and elegance.

However, the restriction is actually not very drastic in view of Whitney's result that, on an n-manifold X, every differentiable structure of class C^k with $k > 0$ contains a differentiable structure of class C^∞, [Mu, p. 53].

Let X denote an arbitrary smooth n-manifold with a given smooth structure σ. Hereafter, every chart $\phi : U \to R^n$ in the maximum atlas μ_σ of the smooth structure σ will be simply called a *smooth chart* of X.

Let x denote any given point of X. By a *local smooth function* at the point x of X, we mean a smooth function

$$f : U \to R$$

defined on an open neighborhood U of x in X. Let

$$L(X,x)$$

denote the set of all local smooth functions at the point x of X.

In the set $L(X,x)$, we can define scalar multiplication, addition, and multiplication as follows. For arbitrary a, $b \in R$ and any two local smooth functions

$$f : U \to R \qquad g : V \to R$$

in $L(X,x)$, we have the local smooth functions

$$af + bg : U \cap V \to R \qquad fg : U \cap V \to R$$

defined by

$$(af + bg)(w) = a[f(w)] + b[g(w)],$$
$$(fg)(w) = f(w)g(w)$$

for every $w \in U \cap V$. However, these operations fail to make $L(X,x)$ an algebra over R. In fact, $L(X,x)$ is not even a linear space over R because we have

$$f + (-f) \neq g + (-g)$$

unless $U = V$.

To remove this undesirable behavior of $L(X,x)$, let us define a relation $\sim$ in the set $L(X,x)$ as follows. For any two local smooth functions

$$f : U \to R \qquad g : V \to R$$

in $L(X,x)$, $f \sim g$ iff there exists an open neighborhood $W \subset U \cap V$ of x in X such that

$$f(w) = g(w)$$

holds for every point w in W. Since this relation in $L(X,x)$ is obviously reflexive, symmetric, and transitive, it is an equivalence relation. Hence $\sim$ divides the members of $L(X,x)$ into disjoint equivalence classes, called the *germs of local smooth functions* at the point x of X. We shall simply call these the *smooth germs* at the point x of X. Let

$$G(X,x) = L(X,x)/\sim$$

denote the set of all smooth germs at the point x of X and let

$$p : L(X,x) \to G(X,x)$$

denote the natural projection of the set $L(X,x)$ onto its quotient set $G(X,x)$.

The following lemma is obvious from the definition of the relation $\sim$ in $L(X,x)$.

Lemma 4.1. *If $f_1 \sim f_2$ and $g_1 \sim g_2$ in $L(X,x)$, then*

$$af_1 + bg_1 \sim af_2 + bg_2 \qquad f_1g_1 \sim f_2g_2$$

for every $a \in R$ and every $b \in R$.

Because of (4.1), we can define scalar multiplication, addition, and multiplication in $G(X,x)$ by taking

$$\begin{aligned} a\omega + b\theta &= p(af + bg), \\ \omega\theta &= p(fg) \end{aligned}$$

for arbitrary smooth germs ω, $\theta \in G(X,x)$ any arbitrary real numbers $a, b \in R$, with $f \in \omega$ and $g \in \theta$. One can easily verify that these operations make $G(X,x)$ an *algebra* over the field R of real numbers, which will be referred to as the *algebra of smooth germs* at the point x of X.

The following proposition is obvious.

Proposition 4.2. *The algebra $G(X,x)$ is associative and commutative. Besides, it has a unit $\iota = p(u)$, where $u : X \to R$ stands for the smooth function defined by $u(y) = 1$ for every point y of X.*

The following lemma is obvious from the definition of the relation $\sim$ in $L(X,x)$.

Lemma 4.3. *If $f \sim g$ in $L(X,x)$, then $f(x) = g(x)$.*

On the basis of (4.3), we may define the value $\omega(x)$ of a smooth germ $\omega \in G(X,x)$ at the point x by taking

$$\omega(x) = f(x)$$

for every local smooth function $f \in \omega$.

Now let us define the tangent vectors.

By a *tangent vector* of X at the point x, we mean a function

$$v : G(X,x) \to R$$

that satisfies the following two conditions:

(TV1) v is a *linear functional* on the linear space $G(X,x)$; that is,

$$v(a\omega + b\theta) = av(\omega) + bv(\theta)$$

holds for arbitrary ω, $\theta \in G(X,x)$ and a, $b \in R$.

(TV2) v is a *derivation;* that is,

$$v(\omega\theta) = v(\omega)\theta(x) + \omega(x)v(\theta)$$

holds for arbitrary ω, $\theta \in G(X,x)$.

We shall use the symbol

$$T(X,x)$$

to denote the set of all tangent vectors of X at the point x. For arbitrary v, $w \in T(X,x)$ and a, $b \in R$, the function

$$av + bw : G(X,x) \to R$$

defined by

$$(av + bw)(\omega) = av(\omega) + bw(\omega)$$

for every $\omega \in G(X,x)$ is clearly in $T(X,x)$. One can easily verify that this makes $T(X,x)$ a *linear space* over the field R of real numbers. This linear space $T(X,x)$ is called the *tangent space* of X at the point x.

Let $K(X,x)$ denote the subset of $G(X,x)$ which consists of all smooth germs $\omega \in G(X,x)$ such that $v(\omega) = 0$ holds for every tangent vector $v \in T(X,x)$. One can easily see that $K(X,x)$ is a subalgebra of the algebra $G(X,x)$. In particular, $K(X,x)$ is a linear subspace of the linear space $G(X,x)$. The quotient linear space

$$T^*(X,x) = G(X,x)/K(X,x)$$

is known as the *cotangent space* of X at the point x. Let

$$q : G(X,x) \to T^*(X,x)$$

denote the natural projection of the linear space $G(X,x)$ onto its quotient linear space $T^*(X,x)$.

This completes the definition of $T(X,x)$ and $T^*(X,x)$. We are going to determine their structures.

For this purpose, let us consider an arbitrary smooth chart

$$\phi : U \to R^n$$

of X with $x \in U$. For every integer $i = 1, 2, \ldots, n$, let

$$\frac{\partial}{\partial \phi_i} \in T(X,x)$$

be defined as follows. For any $\omega \in G(X,x)$, select a local smooth function

$$f : V \to R$$

in $L(X,x)$ with $p(f) = \omega$. Since the function

$$g : \phi(U \cap V) \to R$$

defined by

$$g(t) = f[\phi^{-1}(t)]$$

for every $t \in \phi(U \cap V)$ is a smooth function on the open subspace $\phi(U \cap V)$ of R^n, the partial derivative

$$\left(\frac{\partial g}{\partial t_i}\right)[\phi(x)] \in R$$

at the point $\phi(x) \in \phi(U \cap V)$ is well defined and does not depend on the choice of f from the germ ω. Hence we may define $\partial/\partial\phi_i$ by taking

$$\frac{\partial}{\partial \phi_i}(\omega) = \left(\frac{\partial g}{\partial t_i}\right)[\phi(x)].$$

The verification of

$$\frac{\partial}{\partial \phi_i} \in T(X,x)$$

is straightforward and, hence, is omitted.

We shall prove that the n tangent vectors

$$\frac{\partial}{\partial \phi_1}, \ldots, \frac{\partial}{\partial \phi_n}$$

constitute a basis of the linear space $T(X,x)$. For this purpose, let us first establish the following lemmas.

Lemma 4.4. *The unit ι of the algebra $G(X,x)$ is contained in the subalgebra $K(X,x)$.*

PROOF. Because of $\iota = \iota\iota$, it follows from the condition (TV2) that we have

$$v(\iota) = v(\iota\iota) = v(\iota) + v(\iota)$$

for every $v \in T(X,x)$. This implies $v(\iota) = 0$ for every $v \in T(X,x)$ and hence $\iota \in K(X,x)$.∎

According to Exercise 3A, the coordinate functions

$$\phi_i : U \to R \qquad (i = 1, 2, \ldots, n)$$

in U with respect to the chart $\phi : U \to R^n$ are smooth and hence in $L(X,x)$.

Lemma 4.5. *For every local smooth function $f : V \to R$ in $L(X,x)$, there exists an open neighborhood*

$$W \subset U \cap V$$

of x in X together with n smooth functions

$$f_i : W \to R \qquad (i = 1, 2, \ldots, n)$$

such that

$$f_i(x) = \frac{\partial}{\partial \phi_i} [p(f)]$$

holds for every $i = 1, 2, \ldots, n$ and

$$f(y) = f(x) + \sum_{i=1}^{n} [\phi_i(y) - \phi_i(x)] f_i(y)$$

holds for every $y \in W$.

PROOF. Consider the smooth function

$$g : \phi(U \cap V) \to R$$

defined on the open subspace $\phi(U \cap V)$ of R^n by

$$g(t) = f[\phi^{-1}(t)]$$

for every $t \in \phi(U \cap V)$. Let

$$a = (a_1, \ldots, a_n) = \phi(x) \in R^n.$$

Since $\phi(U \cap V)$ is an open neighborhood of a in R^n, there exists a positive real number δ such that $\phi(U \cap V)$ contains the open δ-ball

$$B_\delta(a) = \{t \in R^n \mid \|t - a\| < \delta\}.$$

Then the partial derivatives

$$\frac{\partial g}{\partial t_i} \qquad (i = 1, 2, \ldots, n)$$

of g are smooth functions in $B_\delta(a)$. For every $i = 1, 2, \ldots, n$, define a smooth function

$$g_i : B_\delta(a) \to R$$

by taking

$$g_i(b) = \int_0^1 \left(\frac{\partial g}{\partial t_i}\right) [b_1, \ . \ . \ . \ , b_{i-1}, a_i + \lambda(b_i - a_i), a_{i+1}, \ . \ . \ . \ , a_n] \, d\lambda$$

for every point $b = (b_1, \ . \ . \ . \ , b_n)$ in $B_\delta(a)$.

Multiplying the last exhibited equality by $b_i - a_i$ and summing over $i = 1, 2, \ . \ . \ . \ , n$, we obtain

$$\begin{aligned} &\sum_{i=1}^{n} (b_i - a_i) g_i(b) \\ &\quad = \sum_{i=1}^{n} (b_i - a_i) \int_0^1 \left(\frac{\partial g}{\partial t_i}\right) [b_1, \ . \ . \ . \ , b_{i-1}, a_i + \lambda(b_i - a_i), a_{i+1}, \ . \ . \ . \ , a_n] \, d\lambda \\ &\quad = \sum_{i=1}^{n} [g(b_1, \ . \ . \ . \ , b_{i-1}, a_i, \ . \ . \ . \ , a_n) - g(b_1, \ . \ . \ . \ , b_i, a_{i+1}, \ . \ . \ . \ , a_n)] \\ &\quad = g(b) - g(a) \end{aligned}$$

for every point $b \in B_\delta(a)$. Hence we have

$$g(b) = g(a) + \sum_{i=1}^{n} (b_i - a_i) g_i(b)$$

for every point $b \in B_\delta(a)$.

Let $W = \phi^{-1}[B_\delta(a)] \subset U \cap V$ and let

$$f_i : W \to R \qquad (i = 1, 2, \ . \ . \ . \ , n)$$

denote the smooth functions defined by

$$f_i(y) = g_i[\phi(y)]$$

for every $y \in W$. Then, by the definition of $\partial/\partial\phi_i$, it is clear that

$$f_i(x) = \frac{\partial}{\partial \phi_i} [p(f)]$$

holds for every $i = 1, 2, \ . \ . \ . \ , n$. On the other hand, let y denote an arbitrary point in W. Then we have

$$b = \phi(y) \in B_\delta(a).$$

Consequently, we obtain

$$\begin{aligned} f(y) &= f[\phi^{-1}(b)] = g(b) \\ &= g(a) + \sum_{i=1}^{n} (b_i - a_i)g_i(b) \\ &= f(x) + \sum_{i=1}^{n} [\phi_i(y) - \phi_i(x)]f_i(y). \end{aligned}$$

This completes the proof of (4.5).||

Lemma 4.6. *For every $v \in T(X,x)$, we have*

$$v = \sum_{i=1}^{n} v[p(\phi_i)] \frac{\partial}{\partial \phi_i}.$$

PROOF. Let $v \in T(X,x)$ and $\omega \in G(X,x)$ be arbitrarily given. Select a local smooth function $f : V \to R$ in $L(X,x)$ with $p(f) = \omega$. According to (4.5), we have

$$\omega = p(f) = f(x)\iota + \sum_{i=1}^{n} [p(\phi_i) - \phi_i(x)\iota]p(f_i)$$

in the algebra $G(X,x)$. By (4.4) and the conditions (TV1)-(TV2), we obtain

$$v(\omega) = \sum_{i=1}^{n} v[p(\phi_i)]f_i(x) = \sum_{i=1}^{n} v[p(\phi_i)] \frac{\partial}{\partial \phi_i}(\omega).$$

Since this holds for every $\omega \in G(X,x)$, we have

$$v = \sum_{i=1}^{n} v[p(\phi_i)] \frac{\partial}{\partial \phi_i}$$

in $T(X,x)$. This completes the proof of (4.6).▌

Lemma 4.7. *The tangent vectors*

$$\frac{\partial}{\partial \phi_1}, \ldots, \frac{\partial}{\partial \phi_n}$$

in $T(X,x)$ are linearly independent.

PROOF. First we observe from the definition of

$$\frac{\partial}{\partial \phi_i}$$

that we have

$$\frac{\partial}{\partial \phi_i}[p(\phi_j)] = \begin{cases} 1 & (\text{if } i = j) \\ 0 & (\text{if } i \neq j). \end{cases}$$

To prove (4.7), let us assume that a certain linear combination

$$v = \sum_{i=1}^{n} a_i \frac{\partial}{\partial \phi_i}$$

is the zero element of $T(X,x)$; that is, $v = 0$. Then, for every $j = 1, 2, \ldots, n$, we have

$$0 = v[p(\phi_j)] = \sum_{i=1}^{n} a_i \frac{\partial}{\partial \phi_i}[p(\phi_j)] = a_j.$$

This completes the proof of (4.7).▌

The following theorem is a direct consequence of (4.6) and (4.7).

Theorem 4.8. *The tangent space $T(X,x)$ of any smooth n-manifold X at an arbitrary point x of X is an n-dimensional linear space over R. For every smooth chart $\phi : U \to R^n$ of X with $x \in U$, the n tangent vectors*

$$\frac{\partial}{\partial \phi_1}, \ldots, \frac{\partial}{\partial \phi_n}$$

constitute a basis of $T(X,x)$.

Next let us determine the structure of the cotangent space $T^*(X,x)$.

For this purpose, let us consider the composition

$$d_x = q \circ p : L(X,x) \to T^*(X,x)$$

of the natural projections $p : L(X,x) \to G(X,x)$ and $q : G(X,x) \to T^*(X,x)$. For every local smooth function

$$f : U \to R$$

in $L(X,x)$, the cotangent vector $d_x(f)$ in $T^*(X,x)$ is called the *differential* of f at the point x.

We observe that the algebra

$$F(X) = \text{Fun}^\infty(X)$$

of all smooth functions on X is a subset of $L(X,x)$. This algebra $F(X)$ over R is commutative and associative. Furthermore, it has a unit $u : X \to R$ defined by $u(y) = 1$ for every point y of X.

Lemma 4.9. $p[F(X)] = G(X,x)$.

PROOF. Let ω denote any smooth germ in $G(X,x)$. Select a local smooth function $f : U \to R$ in the germ ω. Since $x \in U$ and X is locally compact, there exists a compact neighborhood A of x contained in U. As a paracompact Hausdorff space, X is normal. Hence there exists an open neighborhood V of A such that its closure $\mathrm{Cl}(V)$ is contained in U. According to (3.10), there exists a smooth function $g : X \to R$ such that

$$g(A) = 1 \qquad g(X \setminus V) = 0.$$

Define a function $h : X \to R$ by taking

$$h(y) = \begin{cases} f(y)g(y) & \text{if } y \in U \\ 0 & \text{if } y \in X \setminus \mathrm{Cl}(V). \end{cases}$$

Since h is smooth in both U and $X \setminus \mathrm{Cl}(V)$, we have $h \in F(X)$. Since $h(y) = f(y)$ for every $y \in A$, we have $h \sim f$. Hence we obtain

$$p(h) = p(f) = \omega.$$

Since ω is any germ in $G(X,x)$, this proves (4.9).∎

Since the natural projection q is surjective, the following proposition is a direct consequence of (4.9).

Proposition 4.10. $d_x[F(X)] = T^*(X,x)$; *in words, every cotangent vector of X at x is the differential at x of a smooth function $f : X \to R$.*

Although $F(X)$ has better algebraic structure than $L(X,x)$, the latter is frequently more convenient to handle.

Next let us consider m arbitrarily given local smooth functions

$$f_i : U_i \to R \qquad (i = 1, 2, \ldots, m)$$

in $L(X,x)$. For every $i = 1, 2, \ldots, m$, let $a_i = f_i(x) \in R$. Then

$$a = (a_1, a_2, \ldots, a_m)$$

is a point of the Euclidean n-space R^m. Let

$$\Phi : V \to R$$

denote any smooth function defined on an open neighborhood V of the point a in R^m. Then there exists an open neighborhood W of x in X such that

$$W \subset U_1 \cap U_2 \cap \cdots \cap U_m$$

and that the point

$$t = [f_1(w), f_2(w), \ldots, f_m(w)]$$

of R^m is in V whenever w is in W. Define a function

$$f : W \to R$$

by taking

$$f(w) = \Phi(t) = \Phi[f_1(w), f_2(w), \ldots, f_m(w)]$$

for every $w \in W$. Then f is smooth and hence is in $L(X,x_0)$.

Lemma 4.11. *The differential of f at the point x of X is given by*

$$d_x(f) = \sum_{i=1}^{m} \lambda_i d_x(f_i)$$

where λ_i stands for the partial derivative

$$\lambda_i = \left(\frac{\partial \Phi}{\partial t_i}\right)(a)$$

of Φ at the point $a \in V$ for every $i = 1, 2, \ldots, m$.

PROOF. Since $f, f_1, \ldots, f_m$ are in $L(X,x)$,

$$g = f - \sum_{i=1}^{m} \lambda_i f_i$$

is a well-defined local smooth function in $L(X,x)$. Since $d_x = q \circ p$, it suffices to prove that the smooth germ $p(g)$ of $G(X,x)$ is in the subalgebra $K(X,x)$. For this purpose, we have to show that

$$v[p(g)] = 0$$

holds for every tangent vector $v \in T(X,x)$.

Let $\phi : U \to R^n$ denote an arbitrary smooth chart of X with $x \in U$. Then we obtain n tangent vectors

$$\frac{\partial}{\partial \phi_1}, \ldots, \frac{\partial}{\partial \phi_n}$$

in $T(X,x)$. Because of (4.6), it suffices to show that

$$\frac{\partial}{\partial \phi_j}[p(g)] = 0$$

holds for every $j = 1, 2, \ldots, n$. For this purpose, let

$$h, h_1, \ldots, h_m : \phi(U \cap W) \to R$$

denote the smooth functions defined on the open subspace of R^n by

$$h(t) = f[\phi^{-1}(t)] \qquad h_i(t) = f_i[\phi^{-1}(t)]$$

for every $t \in \phi(U \cap W)$ and every $i = 1, 2, \ldots, m$. Let

$$b_j = \left(\frac{\partial h}{\partial t_j}\right)[\phi(x)] \qquad c_{ij} = \left(\frac{\partial h_i}{\partial t_j}\right)[\phi(x)]$$

denote partial derivative at the point $\phi(x) \in R^n$ for every $i = 1, 2, \ldots, m$ and every $j = 1, 2, \ldots, n$.

According to the definition of

$$\frac{\partial}{\partial \phi_j}$$

we have

$$\frac{\partial}{\partial \phi_j}[p(f)] = b_j \qquad \frac{\partial}{\partial \phi_j}[p(f_i)] = c_{ij}.$$

This implies

$$\frac{\partial}{\partial \phi_j}[p(g)] = b_j - \sum_{i=1}^{m} \lambda_i c_{ij}$$

for every $j = 1, 2, \ldots, n$. Since

$$h(t) = \Phi[h_1(t), h_2(t), \ldots, h_m(t)]$$

holds for every $t \in \phi(U \cap W)$, we have

$$b_j = \sum_{i=1}^{m} \lambda_i c_{ij}$$

for every $j = 1, 2, \ldots, n$. This implies

$$\frac{\partial}{\partial \phi_j}[p(g)] = 0$$

for every $j = 1, 2, \ldots, n$ and completes the proof of (4.11).∎

The following corollary is a direct consequence of (4.11) with $m = 2$.

Corollary 4.12. *For any two local smooth functions f, $g \in L(X,x)$ and arbitrary a, $b \in R$, we have*

$$d_x(af + bg) = ad_x(f) + bd_x(g),$$
$$d_x(fg) = g(x)d_x(f) + f(x)d_x(g).$$

By (4.10) and the first equality in (4.12), the restriction $d_x \mid F(X)$ is a linear transformation from $F(X)$ onto $T^*(X,x)$. Hence $T^*(X,x)$ is isomorphic to a quotient of the linear space $F(X)$ of all smooth functions on X.

Now let us consider any smooth chart

$$\phi : U \to R^n$$

of X with $x \in U$. According to Exercise 3A, the n coordinate functions

$$\phi_i : U \to R \qquad (i = 1, 2, \ldots, n)$$

are in $L(X,x)$. For an arbitrary local smooth function

$$f : V \to R$$

in $L(X,x)$, let us consider the smooth function

$$g : \phi(U \cap V) \to R$$

defined on the open neighborhood $\phi(U \cap V)$ of $\phi(x)$ in R^n by

$$g(t) = f[\phi^{-1}(t)]$$

for every $t \in \phi(U \cap V)$. Then we have

$$f(w) = g[\phi_1(w), \phi_2(w), \ldots, \phi_n(w)]$$

for every $w \in U \cap V$. Hence we may apply (4.11) with $m = n$ to this special case and obtain

$$d_x(f) = \sum_{i=1}^{n} \lambda_i d_x(\phi_i)$$

where λ_i stands for the partial derivative

$$\lambda_i = \left(\frac{\partial g}{\partial t_i}\right)[\phi(x)]$$

of g at the point $\phi(x) \in R^n$. On the other hand, we have

$$\frac{\partial}{\partial \phi_i}[p(f)] = \lambda_i$$

according to the definition of $\partial/\partial\phi_i$. Hence we obtain the following corollary of (4.11).

Corollary 4.13. *For every $f \in L(X,x)$, we have*

$$d_x(f) = \sum_{i=1}^{n} \frac{\partial}{\partial \phi_i}[p(f)]\, d_x(\phi_i).$$

Lemma 4.14. *The cotangent vectors $d_x(\phi_1), \ldots, d_x(\phi_n)$ in $T^*(X,x)$ are linearly independent.*

PROOF. Assume that $a_1, \ldots, a_n$ are real numbers such that

$$\sum_{i=1}^{n} a_i d_x(\phi_i) = 0$$

holds in $T^*(X,x)$. Consider the local smooth function

$$f = \sum_{i=1}^{n} a_i \phi_i$$

in $L(X,x)$. Since

$$q[p(f)] = d_x(f) = \sum_{i=1}^{n} a_i d_x(\phi_i) = 0,$$

it follows that $p(f) \in K(X,x)$. This implies $v[p(f)] = 0$ for every $v \in T(X,x)$ In particular, we have

$$\frac{\partial}{\partial \phi_i} [p(f)] = 0$$

for every $i = 1, 2, \ldots, n$. Since

$$\frac{\partial}{\partial \phi_i} [p(\phi_j)] = \begin{cases} 1 & (\text{if } i = j) \\ 0 & (\text{if } i \neq j) \end{cases}$$

we obtain

$$\frac{\partial}{\partial \phi_i} [p(f)] = a_i$$

for every $i = 1, 2, \ldots, n$. This implies $a_i = 0$ for every $i = 1, 2, \ldots, n$ and completes the proof of (4.14).▌

The following theorem is a direct consequence of (4.13) and (4.14).

Theorem 4.15. *The cotangent space $T^*(X,x)$ of any smooth n-manifold X at an arbitrary point x of X is an n-dimensional linear space over R. For every smooth chart $\phi : U \to R^n$ of X with $x \in U$, the n cotangent vectors $d_x(\phi_1), \ldots, d_x(\phi_n)$ constitute a basis of $T^*(X,x)$.*

The remainder of the present section is devoted to the *duality* between $T(X,x)$ and $T^*(X,x)$.

The set $\mathfrak{D}[T(X,x)]$ of all linear functionals on $T(X,x)$ forms a linear space over R in the obvious way. This linear space $\mathfrak{D}[T(X,x)]$ is called the *dual space* of $T(X,x)$. Since $T(X,x)$ is a n-dimensional linear space, so is $\mathfrak{D}[T(X,x)]$.

The following lemma is obvious from the definition of the natural projection

$$q : G(X,x) \to T^*(X,x).$$

Lemma 4.16. *For any two smooth germs $\omega,\theta \in G(X,x)$, we have $q(\omega) = q(\theta)$ iff $v(\omega) = v(\theta)$ holds for every $v \in T(X,x)$.*

Because of (4.16), we may define a function

$$\kappa : T^*(X,x) \to \mathfrak{D}[T(X,x)]$$

by taking

$$[\kappa(\xi)](v) = v(\omega)$$

for every $\xi \in T^*(X,x)$ and every $v \in T(X,x)$, where ω denotes any smooth germ in $G(X,x)$ satisfying $q(\omega) = \xi$.

Lemma 4.17. *κ is linear and bijective.*

PROOF. To prove that κ is linear, let ξ, η be any two cotangent vectors in $T^*(X,x)$ and let a, b denote arbitrary real numbers. Select $\omega,\theta \in G(X,x)$ satisfying $q(\omega) = \xi$ and $q(\theta) = \eta$. Then we have

$$\begin{aligned}[\kappa(a\xi + b\eta)](v) &= v(a\omega + b\theta) \\ &= av(\omega) + bv(\theta) \\ &= a[\kappa(\xi)](v) + b[\kappa(\eta)](v) \\ &= [a\kappa(\xi) + b\kappa(\eta)](v)\end{aligned}$$

for every $v \in T(X,x)$. This implies

$$\kappa(a\xi + b\eta) = a\kappa(\xi) + b\kappa(\eta)$$

and proves that κ is linear.

To prove that κ is bijective, let ξ denote any cotangent vector in $T^*(X,x)$ with $\kappa(\xi) = 0$. Select an $\omega \in G(X,x)$ with $q(\omega) = \xi$. Then we have

$$v(\omega) = [\kappa(\xi)](v) = 0$$

for every $v \in T(X,x)$. This implies $\omega \in K(X,x)$ and hence $\xi = q(\omega) = 0$. Since κ is linear, this implies that κ is injective and that the image $\kappa[T^*(X,x)]$ is an n-dimensional linear subspace of $\mathfrak{D}[T(X,x)]$. Since $\mathfrak{D}[T(X,x)]$ is also n-dimensional, we must have

$$\kappa[T^*(X,x)] = \mathfrak{D}[T(X,x)]$$

and hence κ is also surjective. This proves that κ is bijective.∎

Because of (4.17), κ is a linear isomorphism and will be referred to as the *canonical isomorphism* of $T^*(X,x)$ onto $\mathfrak{D}[T(X,x)]$. By means of κ, we may consider $T^*(X,x)$ as the dual space of $T(X,x)$. Since $T(X,x)$ is finite-dimensional, it can be identified with the dual space of $T^*(X,x)$. Hence, the linear spaces $T(X,x)$ and $T^*(X,x)$ are *dual to each other*. For every $v \in T(X,x)$ and every $T^*(X,x)$, we shall use the notations

$$\langle v,\xi\rangle = v(\xi) = \xi(v)$$

to denote the real number $[\kappa(\xi)](v)$.

Let $\phi : U \to R^n$ denote an arbitrary smooth chart of X with $x \in U$. Then we have

$$\left\langle \frac{\partial}{\partial \phi_i}, d_x(\phi_j) \right\rangle = \begin{cases} 1 & (\text{if } i = j) \\ 0 & (\text{if } i \neq j). \end{cases}$$

Hence

$$\left(\frac{\partial}{\partial \phi_1}, \ . \ . \ . \ , \frac{\partial}{\partial \phi_n}\right) \qquad \text{and} \qquad [d_x(\phi_1), \ . \ . \ . \ , d_x(\phi_n)]$$

are *dual bases.* Thus we have proved the following theorem:

Theorem 4.18. *The tangent space $T(X,x)$ and the cotangent space $T^*(X,x)$ of a smooth n-manifold X at any point x of X are dual to each other. Furthermore, for every smooth chart $\phi : U \to R^n$ of X with $x \in U$,*

$$\left(\frac{\partial}{\partial \phi_1}, \ . \ . \ . \ , \frac{\partial}{\partial \phi_n}\right) \qquad [d_x(\phi_1), \ . \ . \ . \ , d_x(\phi_n)]$$

are dual bases.

EXERCISES

4A. By a *smooth curve* in X at a point $x \in X$, we mean a smooth function $\alpha : U \to X$ defined on an open neighborhood U of 0 in R such that $\alpha(0) = x$. Let $\Gamma(X,x)$ denote the set of all smooth curves in X at x. Define a equivalence relation $\sim$ in $\Gamma(X,x)$ as follows: For any two smooth curves α, $\beta \in \Gamma(X,x)$, $\alpha \sim \beta$ if the derivatives of the composed functions $f \circ \alpha$ and $f \circ \beta$ are equal when $t = 0$ for every $f \in L(X,x)$. Show that the quotient set

$$\Gamma(X,x)/\sim$$

can be identified with the tangent space $T(X,x)$.

4B. Consider the functions $f : R \to R$ and $g : R \to R$ defined for every $t \in R$ by

$$f(t) = \begin{cases} 0 & (\text{if } t \leq 1) \\ e^{-1/(t-1)^2} & (\text{if } t > 1) \end{cases}$$

$$g(t) = 0.$$

Prove that both f and g are smooth functions. At which point of R do f and g belong to the same smooth germ? At which point do they have the same differential?

4C. Consider a smooth curve $\alpha : U \to R^n$ defined on an open neighborhood U of the real number 0 in the real line R. Let

$$f_i : U \to R \qquad (i = 1, 2, \ldots, n)$$

denote the n smooth functions such that

$$\alpha(t) = (f_1(t), f_2(t), \ldots, f_n(t))$$

for every real number t in U. Prove that the tangent vector to α at the point $\alpha(0)$ is

$$\sum_{i=1}^{n} f_i'(0) \frac{\partial}{\partial x_i}.$$

4D. Let X and Y denote arbitrary smooth manifolds. Prove that, for any point (x,y) of the product manifold $X \times Y$, the tangent space $T(X \times Y, (x,y))$ of $X \times Y$ at (x,y) is isomorphic to the direct sum

$$T(X,x) \oplus T(Y,y)$$

of the tangent spaces $T(X,x)$ and $T(Y,y)$. Establish a similar result for cotangent spaces.

4E. Investigate possible generalizations of the concepts and the results given in this section to infinite-dimensional smooth manifolds.

5. Differential of Smooth Map

Throughout the present section, we are concerned with an arbitrary smooth function

$$f : X \to Y$$

from a smooth m-manifold X into a smooth n-manifold Y in the sense of §3. Since f must be continuous, it is usually called a *smooth map* or a *smooth mapping*.

Let x be any point in X and denote

$$y = f(x) \in Y.$$

For every local smooth function $\xi : V \to R$ in $L(Y,y)$, the function $\xi^* : U \to R$ defined on the open neighborhood

$$U = f^{-1}(V)$$

of x in X by

$$\xi^*(w) = \xi[f(w)]$$

for every $w \in U$ is smooth and hence is in $L(X,x)$. The assignment $\xi \rightarrow \xi^*$ defines a function

$$L(f,x) : L(Y,y) \rightarrow L(X,x).$$

The following lemma is obvious.

Lemma 5.1. *For arbitrary* ξ, $\eta \in L(Y,y)$ *and* a, $b \in R$, *the following three statements are true:*

(1) $[L(f,x)](a\xi + b\eta) = a[L(f,x)](\xi) + b[L(f,x)](\eta)$.
(2) $[L(f,x)](\xi\eta) = [L(f,x)](\xi) \cdot [L(f,x)](\eta)$.
(3) $\xi \sim \eta$ *implies* $[L(f,x)](\xi) \sim [L(f,x)](\eta)$.

Because of the statement (3) in (5.1), $L(f,x)$ induces a uniquely determined function

$$f^{\#} = G(f,x) : G(Y,y) \rightarrow G(X,x)$$

such that the rectangle

$$\begin{array}{ccc} L(Y,y) & \xrightarrow{L(f,x)} & L(X,x) \\ \downarrow p & & \downarrow p \\ G(Y,y) & \xrightarrow{G(f,x)} & G(X,x) \end{array}$$

is commutative; that is,

$$p \circ L(f,x) = G(f,x) \circ p$$

holds. Here p stands for the natural projection.

Lemma 5.2. *The function* $G(f,x)$ *is a homomorphism of the algebra* $G(Y,y)$ *into the algebra* $G(X,x)$ *and carries the unit of* $G(Y,y)$ *to the unit of* $G(X,x)$.

PROOF. The first part of (5.2) is an immediate consequence of the statements (1) and (2) in (5.1). The second part of (5.2) is obvious from the definitions of $L(f,x)$ and $G(f,x)$.∎

Furthermore, the following lemma is also obvious from the definition of $f^{\#} = G(f,x)$.

Lemma 5.3. *For every* $\omega \in G(Y,y)$, *we have*

$$[f^{\#}(\omega)](x) = \omega(y).$$

Now let v denote an arbitrary tangent vector in $T(X,x)$. Define a function

$$v_* : G(Y,y) \rightarrow R$$

by taking

$$v_*(\omega) = v[f^\#(\omega)]$$

for every $\omega \in G(Y,y)$.

Lemma 5.4. *v_* is a tangent vector in $T(Y,y)$.*

PROOF. It suffices to verify the conditions (TV1) and (TV2) for v_*. For this purpose, let ω, $\theta \in G(Y,y)$ and a, $b \in R$ be arbitrarily given. Then we have

$$\begin{aligned} v_*(a\omega + b\theta) &= v[f^\#(a\omega + b\theta)] \\ &= v[af^\#(\omega) + bf^\#(\theta)] \\ &= av[f^\#(\omega)] + bv[f^\#(\theta)] \\ &= av_*(\omega) + bv_*(\theta) \\ v_*(\omega\theta) &= v[f^\#(\omega\theta)] \\ &= v[f^\#(\omega)f^\#(\theta)] \\ &= v[f^\#(\omega)] \cdot [f^\#(\theta)](x) + [f^\#(\omega)](x) \cdot v[f^\#(\theta)] \\ &= v_*(\omega)\theta(y) + \omega(y)v_*(\theta) \end{aligned}$$

because of (5.2) and (5.3). This completes the proof of (5.4).▌

Because of (5.4), the assignment $v \to v_*$ defines a function

$$f_* = T(f,x) : T(X,x) \to T(Y,y).$$

Lemma 5.5. *The function $f_* = T(f,x)$ is linear.*

PROOF. Let v, $w \in T(X,x)$ and a, $b \in R$ be arbitrarily given. Then, for every $\omega \in G(Y,y)$, we have

$$\begin{aligned} (av + bw)_*(\omega) &= (av + bw)[f^\#(\omega)] \\ &= av[f^\#(\omega)] + bw[f^\#(\omega)] \\ &= av_*(\omega) + bw_*(\omega) \\ &= (av_* + bw_*)(\omega). \end{aligned}$$

Since this holds for every $\omega \in G(Y,y)$, we obtain

$$f_*(av + bw) = (av + bw)_* = av_* + bw_* = af_*(v) + bf_*(w).$$

This completes the proof of (5.5).▌

This linear transformation $f_* = T(f,x)$ from the tangent space $T(X,x)$ into the tangent space $T(Y,y)$ is called the *differential* of the smooth map f at the point $x \in X$ or the *induced linear transformation on tangent spaces* of the smooth map f at the point $x \in X$.

Lemma 5.6. *$f^\# = G(f,x)$ sends the subalgebra $K(Y,y)$ of $G(Y,y)$ into the subalgebra $K(X,x)$ of $G(X,x)$.*

PROOF. Let $\omega \in K(Y,y)$ be arbitrarily given. To prove $f^{\#}(\omega) \in K(X,x)$, let v denote any tangent vector in $T(X,x)$. According to (5.4), v_* is a tangent vector in $T(Y,y)$. Since $\omega \in K(Y,y)$, this implies $v_*(\omega) = 0$. By the definition of v_*, we have

$$v[f^{\#}(\omega)] = v_*(\omega) = 0.$$

Since this is true for every $v \in T(\mathrm{X},x)$, it follows that $f^{\#}(\omega)$ is in $K(X,x)$. This proves (5.6).▌

Because of (5.6), $f^{\#} = G(f,x)$ induces a uniquely determined linear transformation

$$f^* = T^*(f,x) : T^*(Y,y) \to T^*(X,x)$$

such that the following rectangle is commutative:

$$\begin{array}{ccc} G(Y,y) & \xrightarrow{f\#} & G(X,x) \\ \downarrow q & & \downarrow q \\ T^*(Y,y) & \xrightarrow{f^*} & T^*(X,x) \end{array}$$

where q stands for the natural projections. This linear transformation $f^* = T^*(f,x)$ will be referred to as the *codifferential* of the smooth map f at the point $x \in X$ or the *induced linear transformation on cotangent spaces* of f at $x \in X$.

The following theorem is obvious from the definitions of f_* and f^*.

Theorem 5.7. *The linear transformations f_* and f^* are dual; that is, for every $v \in T(X,x)$ and every $\xi \in T^*(Y,y)$, we have*

$$\langle f_*(v),\ \xi \rangle = \langle v,\ f^*(\xi) \rangle.$$

As dual linear transformations, f_* and f^* determine each other. Hence it suffices to study only one of them, say f_*. The remainder of the present section is devoted to deducing information about the given smooth map f from its differential f_*.

For this purpose, let $\phi : U \to R^m$ denote any smooth chart on X with $x \in U$ and let $\psi : V \to R^n$ denote any smooth chart on Y with $y \in V$. Then the function

$$g : \phi[U \cap f^{-1}(V)] \to R^n$$

defined on the open subspace $\phi[U \cap f^{-1}(V)]$ of R^m by

$$g(t) = \psi\{f[\phi^{-1}(t)]\}$$

for every $t \in \phi[U \cap f^{-1}(V)]$ is smooth. Composing with the natural projections $p_j : R^n \to R$, we obtain n smooth functions

$$g_j = p_j \circ g : \phi[U \cap f^{-1}(V)] \to R \qquad (j = 1, 2, \ldots, n).$$

For each $i = 1, 2, \ldots, m$ and each $j = 1, 2, \ldots, n$, let us consider the partial derivative

$$\lambda_{ij} = \left(\frac{\partial g_j}{\partial t_i}\right)[\phi(x)] \in R$$

at the point $\phi(x)$ of $\phi[U \cap f^{-1}(V)]$. Then the following lemma is an immediate consequence of the chain rule of partial differentiation.

Lemma 5.8. *For every $i = 1, 2, \ldots, m$, we have*

$$f_*\left(\frac{\partial}{\partial \phi_i}\right) = \sum_{j=1}^{n} \lambda_{ij} \frac{\partial}{\partial \psi_j}.$$

Therefore, the $m \times n$ matrix

$$\Lambda = (\lambda_{ij})$$

of real numbers is the matrix of the linear transformation $f_* = T(f,x)$ with respect to the bases

$$\left(\frac{\partial}{\partial \phi_1}, \ldots, \frac{\partial}{\partial \phi_m}\right) \qquad \left(\frac{\partial}{\partial \psi_1}, \ldots, \frac{\partial}{\partial \psi_n}\right)$$

of $T(X,x)$ and $T(Y,y)$. Hence the rank of Λ does not depend on the choice of the smooth charts (ϕ,ψ) and will be called the *rank* of the differential $T(f,x)$.

On the other hand, Λ is the *transpose* of the *Jacobian matrix* $J[g,\phi(x)]$ of the smooth function g at the point $\phi(x)$; in symbols,

$$J[g,\phi(x)] = \Lambda'.$$

Therefore, the rank of $J[g,\phi(x)]$ is equal to that of Λ.

Lemma 5.9. *If $T(f,x)$ is of rank k, then there exist smooth charts $\phi : U \to R^m$ and $\psi : V \to R^n$ on X and Y, respectively, such that $x \in U$, $y \in V$, and*

$$\phi_i(w) = \psi_i[f(w)]$$

for every $i = 1, 2, \ldots, k$ and every $w \in U \cap f^{-1}(V)$.

PROOF. Let $\phi^* : U^* \to R^n$ and $\psi : V \to R^n$ denote arbitrary smooth charts on X and Y respectively such that $x \in U^*$ and $y \in V$. Let

$$g^* : \phi^*[U^* \cap f^{-1}(V)] \to R^n$$

denote the smooth function defined by

$$g^*(t) = \psi\{f[\phi^{*-1}(t)]\}$$

for every $t \in \phi^*[U^* \cap f^{-1}(V)]$ and consider

$$g_j^* = p_j \circ g^* : \phi^*[U^* \cap f^{-1}(V)] \to R \qquad (j = 1, 2, \ldots, n).$$

Since $T(f,x)$ is of rank k, it follows that the $m \times n$ matrix

$$\Lambda^* = (\lambda_{ij}^*) \qquad \lambda_{ij}^* = \left(\frac{\partial g_j^*}{\partial t_i}\right)[\phi^*(x)]$$

is of rank k. Without loss of generality, we may assume that the leading $k \times k$ determinant

$$D = |\lambda_{ij}^*| \qquad (i, j = 1, 2, \ldots, k)$$

is different from zero. This implies the existence of an open neighborhood

$$W \subset \phi^*[U^* \cap f^{-1}(V)]$$

of $\phi^*(x)$ in R^m such that the smooth function $\xi : W \to R^m$, defined by

$$\xi(t) = [g_1^*(t), \ldots, g_k^*(t), t_{k+1}, \ldots, t_m]$$

for every point $t = (t_1, \ldots, t_m)$ of W, is an imbedding.

Let $U = \phi^{*-1}(W)$ and define a function $\phi : U \to R^m$ by taking

$$\phi(w) = \xi[\phi^*(w)]$$

for every $w \in U$. Then $\phi : U \to R^m$ is a smooth chart on X with $x \in U$ and

$$\phi_i(w) = \psi_i[f(w)]$$

for every $i = 1, 2, \ldots, k$ and every $w \in U \cap f^{-1}(V)$. This completes the proof of (5.9).▌

Before deducing consequences of (5.9) let us first define the topological concept of an immersion. By an *immersion* of a topological space X into a topological space Y, we mean a map

$$f : X \to Y$$

such that, for every point $x \in X$, there exists an open neighborhood U of x such that the restriction $f \mid U$ is an imbedding of U into Y.

The following three corollaries are straightforward consequences of (5.9).

Corollary 5.10. *If the differential f_* of a smooth map $f : X \to Y$ is injective for every $x \in X$, then f is an immersion.*

Corollary 5.11. *If the differential f_* of a smooth map $f : X \to Y$ is surjective for every $x \in X$, then f is an open map.*

Corollary 5.12. *If $f : X \to Y$ is a bijective smooth map such that its differential f_* is bijective for every $x \in X$, then f is a diffeomorphism.*

Now let us specialize our study to the smooth identity map

$$f : X \to X$$

on a smooth n-manifold X. Let $\phi : U \to R^n$ and $\psi : V \to R^n$ denote arbitrary smooth charts on X with

$$U \cap V \neq \square.$$

Consider the smooth function

$$g : \phi(U \cap V) \to R^n$$

defined by

$$g(t) = \psi\{f[\phi^{-1}(t)]\} = \psi[\phi^{-1}(t)]$$

for every $t \in \phi(U \cap V)$. The determinant

$$D(\phi,\psi) = \left| \frac{\partial g_i}{\partial t_j} \right| \qquad (i, j = 1, 2, \ldots, n)$$

of the partial derivatives of the n smooth functions

$$g_i = p_i \circ g : \phi(U \cap V) \to R$$

is a real-valued smooth function on the open subspace $\phi(U \cap V)$ of R^n. Since the differential f_* is the identity on $T(X,x)$, we must have

$$[D(\phi,\psi)](t) \neq 0$$

for every $t \in \phi(U \cap V)$.

A smooth n-manifold X is said to be *orientable* iff there exists a smooth atlas α in the smooth structure σ of X such that, for any two smooth charts $\phi : U \to R^n$ and $\psi : V \to R^n$ in α with $U \cap V \neq \square$, we have

$$[D(\phi,\psi)](t) > 0$$

for every $t \in \phi(U \cap V)$. By an *oriented smooth n-manifold*, we mean an n-manifold X together with a smooth atlas α on X such that, for any two charts $\phi : U \to R^n$ and $\psi : V \to R^n$ in α with $U \cap V \neq \square$, we have

$$[D(\phi,\psi)](t) > 0$$

for every $t \in \phi(U \cap V)$. The smooth structure of X shall be the one that contains the given smooth atlas α.

To conclude the present section, we note that, in differential topology and differential geometry, the word "immersion" is often used in the following strong sense. By an *immersion in the strong sense*, we mean a smooth map $f : X \to Y$ such that its differential f_* is injective for every $x \in X$.

EXERCISES

5A. Show that, for any smooth map $f : X \to Y$, the function $L(f,x)$ for any $x \in X$ carries the subset $F(Y)$ of $L(Y,y)$, $y = f(x)$, into the subset $F(X)$ of $L(X,x)$ and defines a homomorphism

$$F(f) : F(Y) \to F(X)$$

from the algebra $F(Y)$ into the algebra $F(X)$.

5B. Establish the following two properties of differentials and codifferentials of smooth maps:

(a) If $i : X \to X$ is the identity map on a smooth manifold X, then $T(i,x)$ and $T^*(i,x)$ are the identity linear transformations on $T(X,x)$ and $T^*(X,x)$ respectively.

(b) If $f : X \to Y$ and $g : Y \to Z$ are smooth maps and $y = f(x)$, then we have

$$\begin{aligned} T(g \circ f,\, x) &= T(g,y) \circ T(f,x) \\ T^*(g \circ f,\, x) &= T^*(f,x) \circ T^*(g,y). \end{aligned}$$

5C. Let $f : X \to Y$ be a smooth map such that $f_* = T(f,x)$ is of rank k for every point x of an open set W of X. By means of (5.9), prove that, for any given point $x \in W$, there exist smooth charts $\phi : U \to R^m$ and $\psi : V \to R^n$ on X and Y, respectively, such that $x \in U$, $y = f(x) \in V$, $\phi(x) = 0$, $\psi(y) = 0$, and

$$g_j(t_1, \ldots, t_m) = \begin{cases} t_j & (\text{if } 1 \leq j \leq k) \\ 0 & (\text{if } k < j \leq n) \end{cases}$$

for every $(t_1, \ldots, t_m) \in \phi[U \cap f^{-1}(V)]$.

5D. Let $f : X \to Y$ be a smooth map of a connected X such that $f_* = T(f,x) = 0$ for every $x \in X$. Prove that f is a constant map.

5E. Every smooth function $f : X \to R$ is also a smooth map of X into the smooth manifold R as defined in Example (a) of §2. Show that, at any point $t \in R$, the tangent space $T(R,t)$ is generated by $\partial/\partial\psi$, where ψ denotes the identity chart $\psi : R \to R$. Prove that, for every $x \in X$, we have

$$[T(f,x)](v) = [d_x(f)](v)\, \frac{\partial}{\partial \psi}$$

for every $v \in T(X,x)$. Hence we may identify $T(f,x)$ with $d_x(f)$ if we have identified $T(R,t)$ with R.

5F. Investigate possible generalizations of the concepts and the results given in this section to infinite-dimensional smooth manifolds.

6. Vector Bundles

Throughout the present section, let X denote an arbitrary smooth n-manifold.

By a *smooth k-dimensional vector bundle over X*, we mean a smooth map

$$\pi : E \to X$$

of a smooth $(n + k)$-manifold E onto X such that the following two conditions are satisfied:

(VB1) For each $x \in X$, the inverse image $\pi^{-1}(x)$ in E has been given the structure of a k-dimensional linear space over R.

(VB2) For each $x \in X$, there exists an open neighborhood U of x in X together with a diffeomorphism

$$\tau : \pi^{-1}(U) \to U \times R^k$$

such that the commutativity relation

$$p \circ \tau = \pi_U$$

holds in the diagram

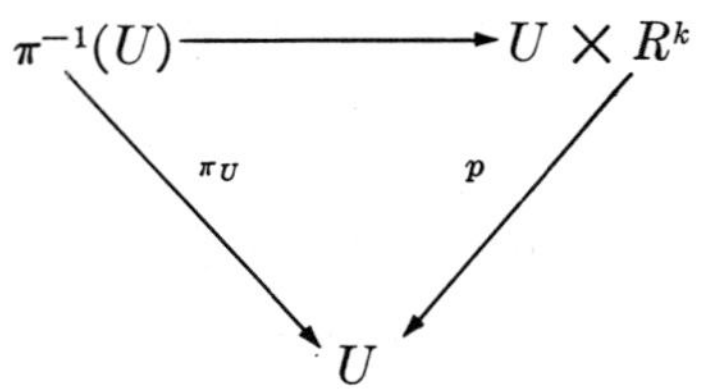

and that the function

$$\tau_x : \pi^{-1}(x) \to R^k$$

defined by

$$\tau_x(e) = q[\tau(e)]$$

for every $e \in \pi^{-1}(x)$ is a bijective linear map.

Here π_U stands for the map defined by π on $\pi^{-1}(U)$ and

$$p : U \times R^k \to U \qquad q : U \times R^k \to R^k$$

denote the natural projections.

Let $\pi : E \to X$ denote any smooth k-dimensional vector bundle over X. We say that E is the *total space* of π and X is the *base space* of π. For each $x \in X$, the subspace $\pi^{-1}(x)$ of E is called the *fiber* over the point x. The open neighborhood U of x in (VB2) is called a *decomposing neighborhood* of x and the diffeomorphism

$$\tau : \pi^{-1}(U) \to U \times R^k$$

will be referred to as the *decomposing function.*

Examples of smooth vector bundles:

(a) *The product bundle* $X \times R^k$. Consider the topological product $X \times R^k$ and the natural projection

$$\pi : X \times R^k \to X.$$

By Example (e) of §2, $X \times R^k$ is a smooth $(n + k)$-manifold and π is clearly a smooth map. For each $x \in X$, the inverse image

$$\pi^{-1}(x) = x \times R^k$$

inherits the linear structure of R^k. Hence (VB1) is satisfied. For (VB2), let us take $U = X$. Then the identity diffeomorphism

$$\tau : \pi^{-1}(U) \to U \times R^k$$

is a decomposing function. Hence π is a smooth $(n + k)$-dimensional vector bundle over X.

(b) *The tangent bundle* $T(X)$. Consider the set

$$T(X) = \bigcup_{x \in X} T(X,x)$$

of all tangent vectors in X. Define a function

$$\pi : T(X) \to X$$

by taking $\pi(v) = x$ for every $v \in T(X,x)$ and every $x \in X$. For each $x \in X$, the inverse image

$$\pi^{-1}(x) = T(X,x)$$

is an n-dimensional linear space over R according to (4.8). Next let

$$\phi : U_\phi \to R^n$$

denote an arbitrary smooth chart on X. According to (4.8),

$$\left(\frac{\partial}{\partial \phi_1}, \ldots, \frac{\partial}{\partial \phi_n}\right)$$

is a basis of $T(X,x)$ for every $x \in U_\phi$. Hence every tangent vector v in $\pi^{-1}(U_\phi)$ can be uniquely expressed by

$$v = \sum_{i=1}^{n} \xi_i(v) \frac{\partial}{\partial \phi_i}$$

where $\xi_1(v), \ldots, \xi_n(v)$ are real numbers that depend on v. Define a function

$$\tau_\phi : \pi^{-1}(U_\phi) \to U_\phi \times R^n$$

by taking

$$\tau_p(v) = \pi(v) \times [\xi_1(v), \ldots, \xi_n(v)]$$

for every $v \in \pi^{-1}(U_\phi)$. Then τ_ϕ is clearly bijective and, therefore, induces a topology in $\pi^{-1}(U_\phi)$ as follows: A subset W of $\pi^{-1}(U_\phi)$ is open iff $\tau_\phi(W)$ is open in $U_\phi \times R^n$. Since

$$T(X) = \bigcup_{\phi \in \mu_\sigma} \pi^{-1}(U_\phi)$$

where μ_σ denotes the maximal smooth atlas of X, we may define a topology in $T(X)$ as follows: A subset W of $T(X)$ is open iff $W \cap \pi^{-1}(U_\phi)$ is open in $\pi^{-1}(U_\phi)$ for every $\phi \in \mu_\sigma$. In this way, $T(X)$ becomes a topological manifold and τ_ϕ becomes a homeomorphism. For every $\phi \in \mu_\sigma$, define a function

$$\phi^* : \pi^{-1}(U_\phi) \to R^{2n} = R^n \times R^n$$

by taking

$$\phi^*(v) = \phi[\pi(v)] \times [\xi_1(v), \ldots, \xi_n(v)]$$

for every $v \in \pi^{-1}(U_\phi)$. Then ϕ^* is a chart on $T(X)$ and the collection

$$\alpha = \{\phi^* \mid \phi \in \mu_\sigma\}$$

is a smooth atlas on $T(X)$. This atlas α makes $T(X)$ a smooth $2n$-manifold. It is straightforward to verify that

$$\pi : T(X) \to X$$

becomes a smooth vector bundle over X with

$$\tau_\phi : \pi^{-1}(U_\phi) \to U_\phi \times R^n$$

as a decomposing function for every $\phi \in \mu_\sigma$. This n-dimensional smooth vector bundle over X is known as the *tangent bundle* of X.

(c) *The cotangent bundle* $T^*(X)$. If, in the construction of the tangent bundle, we replace $T(X,x)$ by $T^*(X,x)$ and $(\partial/\partial\phi_1, \ldots, \partial/\partial\phi_n)$ by $(d\phi_1, \ldots, d\phi_n)$, we obtain an n-dimensional smooth vector bundle

$$\pi^* : T^*(X) \to X$$

over X, known as the *cotangent bundle* of X.

Now let us consider an arbitrary smooth vector bundle

$$\pi : E \to X$$

over the given smooth manifold X.

By a *smooth cross section* of π, we mean a smooth map

$$\gamma : X \to E$$

such that the composition $\pi \circ \gamma$ is the identity map on X. Let

$$\Gamma(\pi)$$

denote the set of all smooth cross sections of π. Since $\pi^{-1}(x)$ is a linear space over R for every $x \in X$, we may define a linear structure in $\Gamma(\pi)$ as follows: Let α, $\beta \in \Gamma(\pi)$ and a, $b \in R$ be arbitrarily given. Then the function

$$a\alpha + b\beta : X \to E$$

defined by

$$(a\alpha + b\beta)(x) = a[\alpha(x)] + b[\beta(x)]$$

for every $x \in X$ is clearly a smooth cross section of π. Hence we have the following proposition:

Proposition 6.1. *$\Gamma(\pi)$ is a linear space over R.*

EXERCISES

6A. Let $f : X \to Y$ be a smooth map and $\pi : E \to Y$ be a k-dimensional smooth vector bundle over Y. Let

$$f^*(E) = \{(x,v) \in X \times E \mid f(x) = \pi(v)\}$$

and let

$$f^*(\pi) : f^*(E) \to X \qquad \pi^*(f) : f^*(E) \to E$$

denote the natural projections. Give $f^*(E)$ a suitable smooth structure and prove that $f^*(\pi)$ is a k-dimensional smooth vector bundle over X and that $\pi^*(f)$ is a smooth map which makes the following diagram commutative:

$$\begin{array}{ccc} f^*(E) & \xrightarrow{\pi^*(f)} & E \\ \Big\downarrow{\scriptstyle f^*(\pi)} & & \Big\downarrow{\scriptstyle \pi} \\ X & \xrightarrow{f} & Y \end{array}$$

6B. Let $\pi : E \to X$ denote a k-dimensional smooth vector bundle over X and $\rho : F \to X$ denote an l-dimensional smooth vector bundle over X. Let

$$E \oplus F = \{(u,v) \in E \times F \mid \pi(u) = \rho(v)\}$$

and define a function

$$\pi \oplus \rho : E \oplus F \to X$$

by taking

$$(\pi \oplus \rho)(u,v) = \pi(u) = \rho(v)$$

for every $(u,v) \in E \oplus F$. Give $E \oplus F$ a suitable smooth structure and prove that $\pi \oplus \rho$ is a $(k + l)$-dimensional smooth vector bundle over X, which is known as the *Whitney sum* of π and ρ.

6C. Investigate possible generalizations of the concepts and the results given in this section to possibly infinite-dimensional smooth vector bundles over possibly infinite-dimensional smooth manifolds.

CHAPTER II

Differential Forms

In terms of Grassmann algebra and smooth vector fields, we introduce the notion of smooth differential forms in the first two sections. After establishing the existence of a uniquely determined exterior differentiation, we define the de Rham cohomology groups in the manner as usual in algebraic topology. Induced homomorphisms of a smooth map are carefully presented and the invariance of the de Rham cohomology under diffeomorphisms is proved. The chapter concludes with the important lemma of Poincaré.

1. Grassmann Algebras

Throughout the present section, let V denote an arbitrary n-dimensional linear space over the field R of all real numbers. For any positive integer k, let

$$V^k = V \oplus \cdots \oplus V$$

of k linear spaces, each of which is identical to V.

By an *alternate k-linear function* on V, we mean a function

$$f : V^k \to R$$

satisfying the following conditions:

(AL1) For every $i = 1, 2, \ldots, k$, f is linear in the ith variable. In other words, for arbitrary $v_1, \ldots, v_k, w_i \in V$ and $a,b \in R$, we have

$$\begin{aligned} f(v_1, \ldots, v_{i-1}, av_i + bw_i, v_{i+1}, \ldots, v_k) \\ = af(v_1, \ldots, v_{i-1}, v_i, v_{i+1}, \ldots, v_k) \\ + bf(v_1, \ldots, v_{i-1}, w_i, v_{i+1}, \ldots, v_k). \end{aligned}$$

(AL2) For every permutation π of the k integers $1, 2, \ldots, k$, we have

$$f[v_{\pi(1)}, \ldots, v_{\pi(k)}] = \operatorname{sgn}(\pi)\, f(v_1, \ldots, v_k)$$

for every element $(v_1, \ldots, v_k)$ of V^k.

Here $\operatorname{sgn}(\pi)$ is defined by

$$\operatorname{sgn}(\pi) = \begin{cases} 1 & \text{(if } \pi \text{ is an even permutation)} \\ -1 & \text{(if } \pi \text{ is an odd permutation).} \end{cases}$$

We shall use the symbol

$$\Lambda^k(V)$$

to denote the set of all alternate k-linear functions on the given linear space V.

For any two alternate k-linear functions $f,g \in \Lambda^k(V)$ and any two real numbers a, $b \in R$, the function

$$af + bg : V^k \to R$$

defined by

$$(af + bg)(x) = a[f(x)] + b[g(x)]$$

for every $x \in V^k$ obviously satisfies the two conditions (AL1) and (AL2). Hence we have the following proposition:

Proposition 1.1. *For every $k \geq 1$, $\Lambda^k(V)$ is a linear space over R.*

To determine the dimension of the linear space $\Lambda^k(V)$, let $k \leq n$ and let

$$(e_1, \ldots, e_n)$$

denote any given basis of V. Let

$$c = (e_{i_1}, \ . \ . \ . \ , e_{i_k}) \qquad 1 \leq i_1 < \cdot \cdot \cdot < i_k \leq n$$

denote an arbitrary k-combination of the n elements $(e_1, \ . \ . \ . \ , e_n)$. Because of the conditions (AL1) and (AL2), we can define an alternate k-linear function

$$f_c : V^k \to R$$

by taking

$$f_c(e_{i_1}, \ . \ . \ . \ , e_{i_k}) = 1$$

and

$$f_c(e_{j_1}, \ . \ . \ . \ , e_{j_k}) = 0$$

for every k-tuple $(e_{j_1}, \ . \ . \ . \ , e_{j_k})$ of the set $(e_1, \ . \ . \ . \ , e_n)$ that is not a permutation of c.

By means of the conditions (AL1) and (AL2), one can easily see that

$$\{f_c \mid c \in C^n_k\}$$

where C^n_k stands for the set of all k-combinations of $(e_1, \ . \ . \ . \ , e_n)$, is a basis of the linear space $\Lambda^k(V)$ over R. Hence we obtain the following proposition:

Proposition 1.2. *The dimension of the linear space $\Lambda^k(V)$ over R is equal to the binomial coefficient $\binom{n}{k}$.*

This completes the determination of the structure of the linear spaces $\Lambda^k(V)$ for all $k \leq n$. In particular, since the condition (AL2) becomes trivial when $k = 1$, we have the following proposition:

Proposition 1.3. *$\Lambda^1(V)$ is the dual linear space V^* of all linear functionals on V.*

On the other hand, the following proposition is a straightforward consequence of the conditions (AL1) and (AL2).

Proposition 1.4. $\Lambda^k(V) = 0$ *if* $k > n$.

For the sake of completeness, we also define

$$\Lambda^0(V) = R.$$

Next let us consider the *direct sum*

$$\Lambda(V) = \sum_{k=0}^{n} \Lambda^k(V).$$

Then the following proposition is obvious.

Proposition 1.5. *$\Lambda(V)$ is a linear space over R of dimension 2^n.*

For arbitrarily given $f \in \Lambda^k(V)$ and $g \in \Lambda^l(V)$, let us define a function

$$f \wedge g : V^{k+l} \to R$$

as follows. If $k = 0$ or $l = 0$, $f \wedge g$ is defined by scalar multiplication; otherwise, $f \wedge g$ is defined by taking

$$(f \wedge g)(v) = \frac{1}{(k+l)!} \sum_{\pi \in S_{k+l}} \operatorname{sgn}(\pi) f[v_{\pi(1)}, \ldots, v_{\pi(k)}] g[v_{\pi(k+1)}, \ldots, v_{\pi(k+l)}]$$

for every element $v = (v_1, \ldots, v_{k+l})$ of V^{k+l}. Here S_{k+l} denotes the set of all permutations of the $k + l$ integers $1, 2, \ldots, k + l$. One can easily verify that $f \wedge g$ satisfies the conditions (AL1)–(AL2) and hence is an element of $\Lambda^{k+l}(V)$.

Together with bilinearity, the assignment $(f,g) \to f \wedge g$ defines a bilinear operation

$$\wedge : \Lambda(V) \times \Lambda(V) \to \Lambda(V)$$

which will be referred to as the *exterior multiplication* in $\Lambda(V)$. The following proposition is a straightforward consequence of the definition.

Proposition 1.6. *The exterior multiplication $\wedge$ in $\Lambda(V)$ has the following properties:*

(a) *$\wedge$ is associative.*
(b) *$\wedge$ has $1 \in R = \Lambda^0(V)$ as a two-sided unit; that is,*

$$1 \wedge x = x = x \wedge 1$$

holds for every element x of $\Lambda(V)$.
(c) *$\wedge$ is not commutative, but we have*

$$g \wedge f = (-1)^{kl} f \wedge g$$

for every $f \in \Lambda^k(V)$ and every $g \in \Lambda^l(V)$.

Therefore, the exterior multiplication $\wedge$ makes the linear space $\Lambda(V)$ an algebra over R. Following [C, p. 145], we shall call this algebra $\Lambda(V)$ over R the *Grassmann algebra* of the given linear space V. Because of the direct sum decomposition

$$\Lambda(V) = \sum_{k=0}^{n} \Lambda^k(V)$$

the Grassmann algebra $\Lambda(V)$ is said to be a *graded algebra* over R. For every $k = 0, 1, \ldots, n$, the elements of $\Lambda^k(V)$ are called the *homogeneous elements of degree k of $\Lambda(V)$.*

According to (1.3), the dual space

$$V^* = \Lambda^1(V)$$

is a direct summand of $\Lambda(V)$. Since V is n-dimensional, so is V^*. The following proposition is a direct consequence of the definitions.

Proposition 1.7. *If $\phi_1, \ldots, \phi_n$ is a basis of the dual space V^* of V, then the $\binom{n}{k}$ elements*

$$\phi_{i_1} \wedge \cdots \wedge \phi_{i_k} \qquad 1 \leq i_1 < \cdots < i_k \leq n$$

of $\Lambda^k(V)$ constitute a basis of the linear space $\Lambda^k(V)$ and

$$(\phi_{i_1} \wedge \cdots \wedge \phi_{i_k})(v) = \frac{1}{k!} \sum_{\pi \in S_k} \operatorname{sgn}(\pi) \phi_{i_1}(v_{\pi(1)}) \cdots \phi_{i_k}(v_{\pi(k)})$$

holds for every element $v = (v_1, \ldots, v_k)$ of V^k.

Consequently, the $n + 1$ elements

$$1, \phi_1, \ldots, \phi_n$$

of $\Lambda(V)$ generate the algebra $\Lambda(V)$, while the $2^n - 1$ elements

$$\phi_{i_1} \wedge \cdots \wedge \phi_{i_k} \qquad 1 \leq i_1 < \cdots < i_k \leq n \qquad k = 1, 2, \ldots, n$$

together with $1 \in R = \Lambda^0(V)$, constitute a basis of the linear space $\Lambda(V)$.

Lemma 1.8. *For every $\phi \in V^* = \Lambda^1(V)$, we have $\phi \wedge \phi = 0$.*

PROOF. According to (c) of (1.6),

$$\phi \wedge \phi = (-1)\phi \wedge \phi.$$

This implies $2(\phi \wedge \phi) = 0$ and hence $\phi \wedge \phi = 0$. ▌

By means of (1.8), we are able to compare the Grassmann algebra $\Lambda(V)$ with the *exterior algebra* $E_R(V^*)$ over R of the dual space V^* defined as follows:

By an *exterior algebra* on a linear space V^* over R, we mean an associative algebra E over R with an identity 1 together with a linear transformation

$$f : V^* \to E$$

which satisfies the following two conditions:

(EA1) $[f(x)]^2 = 0$ holds for every $x \in V^*$.
(EA2) For every linear transformation

$$g : V^* \to X$$

from V^* into an associative algebra X over R with an identity that satisfies

$$[g(x)]^2 = 0$$

for every $x \in V^*$, there exists a unique algebra homomorphism

$$h : E \to X$$

such that $h(1)$ is the identity of X and that

$$h \circ f = g.$$

For any linear space V^* over R, there is an essentially unique exterior algebra on V^* that will be called the *exterior algebra* over R of the linear space V^* and will be denoted by the symbol $E_R(V^*)$.

Now let V^* denote the dual space of the given linear space V. According to the condition (EA2), the inclusion function

$$i : V^* \to \Lambda(V)$$

extends to a unique homomorphism

$$h : E_R(V^*) \to \Lambda(V)$$

of the algebra $E_R(V^*)$ into the algebra $\Lambda(V)$ with

$$h(1) = 1.$$

Proposition 1.9. *$h : E_R(V^*) \to \Lambda(V)$ is an isomorphism.*

PROOF. Since the algebra $\Lambda(V)$ is generated by 1 and V^*, h must be an epimorphism. On the other hand, since the linear spaces $E_R(V^*)$ and $\Lambda(V)$ are both of finite dimension 2^n, the surjective linear transformation h must be bijective. This proves that h is an isomorphism.▌

By means of this isomorphism h, we may identify the Grassmann algebra $\Lambda(V)$ with the exterior algebra $E_R(V^*)$ of its dual space V^*.

Note that we have differentiated the meaning of the terms "Grassmann algebra" and "exterior algebra," which are often used interchangeably in the existing literature.

Now let us consider an arbitrary linear transformation

$$\xi : U \to V$$

from an m-dimensional linear space U into an n-dimensional linear space V. For every $k \geq 1$, define a function

$$\Lambda^k(\xi) : \Lambda^k(V) \to \Lambda^k(U)$$

by taking

$$\{[\Lambda^k(\xi)](f)\}(u_1, \ldots, u_k) = f[\xi(u_1), \ldots, \xi(u_k)]$$

for every $f \in \Lambda^k(V)$ and every $(u_1, \ldots, u_k) \in U^k$. The following proposition is obvious.

Proposition 1.10. *$\Lambda^k(\xi)$ is a linear transformation.*

The following proposition is clear from the definition of $\Lambda^1(\xi)$.

Proposition 1.11. *$\Lambda^1(\xi)$ is the dual linear transformation $\xi^* : V^* \to U^*$ of ξ.*

For completeness, let

$$\Lambda^0(\xi) : \Lambda^0(V) \to \Lambda^0(U)$$

denote the identity linear transformation $i : R \to R$.

Passing to direct sums, we obtain a linear transformation

$$\Lambda(\xi) : \Lambda(V) \to \Lambda(U).$$

The following proposition is a direct consequence of the definitions involved.

Proposition 1.12. *The linear transformation $\Lambda(\xi)$ has the following properties:*

(a) *$\Lambda(\xi)$ preserves degree; that is, it carries $\Lambda^k(V)$ into $\Lambda^k(U)$ for every $k \geq 0$.*

(b) *$\Lambda(\xi)$ preserves unit; that is, $[\Lambda(\xi)](1) = 1$.*

(c) *$\Lambda(\xi)$ preserves exterior products; that is, for arbitrary $f, g \in \Lambda(V)$, we have*

$$[\Lambda(\xi)](f \wedge g) = [\Lambda(\xi)](f) \wedge [\Lambda(\xi)](g).$$

Because of (c) in (1.12), $\Lambda(\xi)$ is a homomorphism of the algebra $\Lambda(V)$ into the algebra $\Lambda(U)$ and will be referred to as the *induced homomorphism* of ξ on the Grassmann algebra. Since

$$\Lambda(V) = E_R(V^*)$$

$\Lambda(\xi)$ is the unique homomorphism from the algebra $\Lambda(V)$ into the algebra $\Lambda(U)$ that reduces to the identity on $\Lambda^0(V)$ and to the dual ξ^* on $\Lambda^1(V)$.

Finally, the functorial properties in the following propositions are obvious.

Proposition 1.13. *If $\xi : V \to V$ is the identity linear transformation, then $\Lambda(\xi)$ is the identity homomorphism on the Grassmann algebra $\Lambda(V)$.*

Proposition 1.14. *If $\xi : U \to V$ and $\eta : V \to W$ are linear transformations, then*

$$\Lambda(\eta \circ \xi) = \Lambda(\xi) \circ \Lambda(\eta).$$

EXERCISES

1A. Show that the Grassmann algebra $\Lambda(V)$ of an n-dimensional linear space V over R can be identified with the dual of the exterior algebra $E_R(V)$ of V.

1B. Let $\xi : V \to V$ denote an arbitrary linear transformation of an n-dimensional linear space V into itself. Prove that

$$[\Lambda^n(\xi)](x) = \Delta(\xi)x$$

holds for every element x of $\Lambda^n(V)$. Here $\Delta(\xi)$ stands for the *determinant* of the linear transformation ξ with respect to any basis $(e_1, \ldots, e_n)$ of V. Hence $\Delta(\xi)$ does not depend on the particular choice of the basis $(e_1, \ldots, e_n)$ of V.

1C. An element x of the Grassmann algebra $\Lambda(V)$ is said to be *homogeneous of degree k iff* $x \in \Lambda^k(V)$. A homogeneous element x of degree $k \geq 1$ of $\Lambda(V)$ is said to be *decomposable* iff there are k elements $d_1, d_2, \ldots, d_n$ of $V^* = \Lambda^1(V)$ such that

$$x = d_1 \wedge d_2 \wedge \cdots \wedge d_k.$$

Establish the following two statements:

(a) If $\dim(V) = n \leq 3$, then every homogeneous element of degree $k \geq 1$ of $\Lambda(V)$ is decomposable.

(b) If $\dim(V) = n > 3$, then

$$(d_1 \wedge d_2) + (d_3 \wedge d_4)$$

is not decomposable for any four linearly independent elements d_1, d_2, d_3, and d_4 of $V^* = \Lambda^1(V)$.

1D. Investigate possible generalizations of the concepts and the results given in this section to the case in which V is an infinite-dimensional linear space over R.

2. Differential Forms

Throughout the present section, let X denote an arbitrary smooth n-manifold. Consider the set

$$T(X) = \bigcup_{x \in X} T(X,x)$$

of all tangent vectors in X together with the *projection*

$$\pi : T(X) \to X$$

defined by taking $\pi(v) = x$ for every $v \in T(X,x)$.

According to Example (b) of (I, §6), $T(X)$ is a smooth $2n$-manifold and π is an n-dimensional smooth vector bundle on X, called the *tangent bundle* of X. However, we shall deliberately avoid the concept of vector bundles in the definition of differential forms (except in identifying remarks) because its use is not absolutely necessary.

By a *vector field* on X, we mean a function

$$V : X \to T(X)$$

such that the composition $\pi \circ V$ of V and the projection $\pi : T(X) \to X$ is the identity function on X.

For any $v \in T(X,x)$ and any $f \in L(X,x)$, we shall use the simpler notation

$$v(f) = v[p(f)]$$

where $p : L(X,x) \to G(X,x)$ denotes the natural projection defined in (I, §4).

A vector field $V : X \to T(X)$ on X is said to be *smooth* iff, for every smooth function $f : X \to R$, the function

$$f_V : X \to R$$

defined by

$$f_V(x) = [V(x)](f)$$

for every $x \in X$, is smooth. Note that we have $f \in L(X,x)$ for every $x \in X$.

One can easily see that the smooth vector fields on X are precisely the smooth cross sections of the tangent bundle

$$\pi : T(X) \to X$$

of X defined in Example (b) of (I, §6) and hence constitute a linear space $\Gamma(\pi)$ over R according to (I,6.1).

Now let us consider an arbitrary smooth chart

$$\phi : U \to R^n$$

on the smooth manifold X. According to Example (c) of (I, §2), U is a smooth manifold and $T(U,x)$ can be clearly identified with $T(X,x)$ for every $x \in U$. Thus we have

$$T(U) = \pi^{-1}(U) = \{T(X,x) \mid x \in U\}.$$

According to (I,4.8), the n tangent vectors

$$\frac{\partial}{\partial \phi_1}, \ldots, \frac{\partial}{\partial \phi_n}$$

in $T(X,x)$ constitute a basis of $T(X,x)$ for every $x \in U$. Hence these are in fact n vector fields on U.

Lemma 2.1. *For every smooth chart $\phi : U \to R^n$ on a smooth manifold X,*

$$\frac{\partial}{\partial \phi_1}, \ . \ . \ . \ , \frac{\partial}{\partial \phi_n}$$

are smooth vector fields on U.

PROOF. It remains to prove their smoothness. Let $f : U \to R$ denote any smooth function on U. We have to prove that the function

$$f_i = f_{\partial/\partial\phi_i} = \frac{\partial}{\partial \phi_i}(f) : U \to R$$

is also smooth. For this purpose, let us define a function

$$g : \phi(U) \to R$$

on the open subspace $\phi(U)$ by taking

$$g(t) = f[\phi^{-1}(t)]$$

for every point $t = (t_1, \ . \ . \ . \ , t_n)$ of $\phi(U)$. Then g is smooth and we have

$$f_i(x) = \left(\frac{\partial g}{\partial t_i}\right)[\phi(x)] = \left[\left(\frac{\partial g}{\partial t_i}\right) \circ \phi\right](x)$$

for every $x \in U$. As the composition of two smooth functions, f_i is smooth for every $i = 1, 2, \ . \ . \ . \ n$. This completes the proof of (2.1).∎

Now let $v : U \to T(U)$ denote smooth vector field on U. Consider an arbitrary point x of U. Since

$$\frac{\partial}{\partial \phi_1}, \ . \ . \ . \ , \frac{\partial}{\partial \phi_n}$$

is a basis of $T(X,x)$, there exist n real numbers $a_1(x), \ . \ . \ . \ , a_n(x)$ satisfying

$$v(x) = \sum_{i=1}^{n} a_i(x) \frac{\partial}{\partial \phi_i}.$$

Since x may be any point of U, the assignment $x \to a_j(x)$ defines a function

$$a_j : U \to R$$

for every $j = 1, 2, \ . \ . \ . \ , n$.

Lemma 2.2. *For every $j = 1, 2, \ldots, n$, the function $a_j : U \to R$ is smooth.*

PROOF. According to (I, Exercise 3A), the coordinate function

$$\phi_j : U \to R$$

is smooth. Since

$$\frac{\partial}{\partial \phi_i}(\phi_j) = \begin{cases} 1 & (\text{if } i = j) \\ 0 & (\text{if } i \neq j) \end{cases}$$

holds for every $x \in U$, we have

$$v(\phi_j) = \sum_{i=1}^{n} a_i \frac{\partial}{\partial \phi_i}(\phi_j) = a_j.$$

Since v is a smooth vector field on U, this implies that a_j is smooth for every $j = 1, 2, \ldots, n$. ∎

Now let us define the concept of differential forms on X.

For every point $x \in X$, let us consider the Grassmann algebra $\Lambda[T(X,x)]$ and the vector spaces $\Lambda^k[T(X,x)]$, $k = 0, 1, \ldots, n$, of the tangent space $T(X,x)$ as defined in the preceding section. Let

$$\Lambda(X) = \bigcup_{x \in X} \Lambda[T(X,x)]$$

and

$$\Lambda^k(X) = \bigcup_{x \in X} \Lambda^k[T(X,x)]$$

for every $k = 0, 1, \ldots, n$. Besides, let

$$\pi : \Lambda(X) \to X \qquad \pi^k : \Lambda^k(X) \to X$$

denote the projections defined in the obvious way.

As in Examples (b) and (c) of (I, §6), one can give topology and smooth structure to each of $\Lambda(X)$ and $\Lambda^k(X)$ in such a way that π and π^k become smooth vector bundles over X. As we remarked before, we shall not make use of this fact.

By a *differential form* on X, we mean a function

$$\xi : X \to \Lambda(X)$$

such that the composition $\pi \circ \xi$ of ξ and the projection $\pi : \Lambda(X) \to X$ is the identity function on X. Similarly, for any integer $k = 0, 1, \ldots, n$, a *k-form* on X is a function

$$\xi : X \to \Lambda^k(X)$$

such that the composition $\pi^k \circ \xi$ of ξ and the projection $\pi^k : \Lambda^k(X) \to X$ is the identity function on X.

For every $k = 0, 1, \ldots, n$, let $\mathbf{F}^k(X)$ denote the set of all k-forms on X. It has a linear structure over R defined by

$$(a\xi + b\eta)(x) = a[\xi(x)] + b[\eta(x)]$$

for each $x \in X$ and arbitrarily given $\xi, \eta \in \mathbf{F}^k(X)$ and $a, b \in R$. Hence $\mathbf{F}^k(X)$ is a linear space over R. In particular, $\mathbf{F}^0(X)$ is the linear space of all functions

$$\xi : X \to R.$$

Let k denote any integer with $1 \leq k \leq n$. A k-form

$$\xi : X \to \Lambda^k(X)$$

is said to be *smooth* iff, for any k-tuple

$$\sigma = (V_1, V_2, \ldots, V_k)$$

of smooth vector fields $V_1, V_2, \ldots, V_k$ on X, the function

$$\xi_\sigma : X \to R,$$

defined by

$$\xi_\sigma(x) = \xi(x)[V_1(x), V_2(x), \ldots, V_k(x)]$$

for every $x \in X$, is smooth. On the other hand, a 0-form on X is *smooth* iff it is a smooth function on X.

Obviously, the smooth k-forms on X constitute a linear subspace of the linear space $\mathbf{F}^k(X)$, which will be denoted by $\mathbf{C}^k(X)$. In particular, $\mathbf{C}^0(X)$ is the linear space of all smooth functions on X.

Similarly, the set $\mathbf{F}^*(X)$ of all differential forms on X is a linear space over R defined by means of the linear structure of $\Lambda[T(X,x)]$. Because of the direct sum decomposition

$$\Lambda[T(X,x)] = \sum_{k=0}^{n} \Lambda^k[T(X,x)]$$

at every $x \in X$, we have a direct sum decomposition

$$\mathbf{F}^*(X) = \sum_{k=0}^{n} \mathbf{F}^k(X).$$

Hence we obtain a linear subspace

$$\mathbf{C}^*(X) = \sum_{k=0}^{n} \mathbf{C}^k(X)$$

of $\mathbf{F}^*(X)$. The members of $\mathbf{C}^*(X)$ are called the *smooth differential forms* on X.

By means of the exterior multiplication $\wedge$ in $\Lambda[T(X,x)]$ for every $x \in X$, we may describe a bilinear operation

$$\wedge : \mathbf{F}^*(X) \times \mathbf{F}^*(X) \to \mathbf{F}^*(X)$$

defined by

$$(\xi \wedge \eta)(x) = [\xi(x)] \wedge [\eta(x)]$$

for each $x \in X$ and arbitrarily given $\xi, \eta \in \mathbf{F}^*(X)$. This bilinear operation in $\mathbf{F}^*(X)$ will be referred to as the *exterior multiplication*. The following proposition is obvious.

Proposition 2.3. *The exterior multiplication $\wedge$ in $\mathbf{F}^*(X)$ makes $\mathbf{F}^*(X)$ an associative algebra over R with the following properties:*

(a) *The constant function $1 \in \mathbf{C}^0(X)$ is a two-sided multiplicative unit, that is,*

$$1 \wedge \xi = \xi = \xi \wedge 1$$

holds for every element ξ of $\mathbf{F}^*(X)$.

(b) *The linear subspace $\mathbf{C}^*(X)$ of $\mathbf{F}^*(X)$ is a subalgebra of $\mathbf{F}^*(X)$; that is,*

$$\xi \wedge \eta \in \mathbf{C}^*(X)$$

for any two smooth differential forms ξ and η on X.

(c) *For arbitrary $\xi \in \mathbf{F}^k(X)$ and $\eta \in \mathbf{F}^l(X)$,*

$$\eta \wedge \xi = (-1)^{kl}\xi \wedge \eta \in \mathbf{F}^{k+l}(X).$$

One can easily verify that the smooth differential forms on X are precisely the smooth cross sections of the smooth vector bundle

$$\pi : \Lambda(X) \to X$$

and that the smooth k-forms on X are precisely the smooth cross sections of the smooth vector bundle

$$\pi^k : \Lambda^k(X) \to X.$$

Now let us consider an arbitrary smooth chart

$$\phi : U \to R^n$$

on X. Then U is a smooth manifold and $T(U,x)$ can be identified with $T(X,x)$ for every $x \in U$. Thus we have

$$\Lambda(U) = \pi^{-1}(U) = \{\Lambda[T(X,x)] \mid x \in U\}.$$

Proposition 2.4. *For an arbitrary k-form*

$$\xi : U \to \Lambda^k(U)$$

on U, there exist $\binom{n}{k}$ uniquely determined functions

$$\xi_{(i_1, \ldots, i_k)} : U \to R \qquad 1 \leq i_1 < \cdots < i_k \leq n$$

such that

$$\xi(x) = \Sigma \xi_{(i_1, \ldots, i_k)}(x)[d_x(\phi_{i_1}) \wedge \cdots \wedge d_x(\phi_{i_k})]$$

holds for every $x \in U$. Furthermore, ξ is smooth iff these $\binom{n}{k}$ functions $\xi_{(i_1, \ldots, i_k)}$ are smooth.

PROOF. According to (I,4.15), the n cotangent vectors

$$d_x(\phi_1), \ldots, d_x(\phi_n)$$

constitute a basis of the cotangent space $T^*(X,x)$ for every $x \in U$. Hence it follows from (1.7) that the $\binom{n}{k}$ elements

$$d_x(\phi_{i_1}) \wedge \cdots \wedge d_x(\phi_{i_k}) \qquad 1 \leq i_1 < \cdots < i_k \leq n$$

of $\Lambda^k[T(X,x)]$ constitute a basis of the linear space $\Lambda^k[T(X,x)]$. Since $\xi(x)$ is an element of $\Lambda^k[T(X,x)]$, there exist $\binom{n}{k}$ uniquely determined real numbers

$$\xi_{(i_1, \ldots, i_k)}(x) \qquad 1 \leq i_1 < \cdots < i_k \leq n$$

such that

$$\xi(x) = \Sigma \xi_{(i_1, \ldots, i_k)}(x)[d_x(\phi_{i_1}) \wedge \cdots \wedge d_x(\phi_{i_k})]$$

where the summation is taken over the set C^n_k of all combinations $(i_1, \ldots, i_k)$ of the integers $1, 2, \ldots, n$. Since this is true for all $x \in U$, we obtain the $\binom{n}{k}$ uniquely determined functions $\xi_{(i_1, \ldots, i_k)}$ and the first part of (2.4).

To establish the necessity of the second part of (2.4), let us assume that ξ is smooth. Consider an arbitrary combination

$$(i_1, \ldots, i_k), \qquad 1 \leq i_1 < \cdots < i_k \leq n.$$

According to (2.1), we have a k-tuple

$$\sigma = (V_1, \ldots, V_k)$$

of smooth vector fields

$$V_j = \frac{\partial}{\partial \phi_{i_j}} \qquad (j = 1, 2, \ldots, k)$$

on U. By a straightforward application of (1.7) and (I,4.8), we obtain

$$\xi_\sigma(x) = \xi(x)[V_1(x), \ldots, V_k(x)] = \frac{1}{k!} \xi_{(i_1, \ldots, i_k)}(x)$$

for every $x \in U$. Therefore,

$$\xi_{(i_1, \ldots, i_k)} = (k!)\xi_\sigma.$$

By the definition of the smoothness of ξ, ξ_σ is smooth and hence so is $\xi_{(i_1,\ldots,i_k)}$. This establishes the necessity of the second part of (2.4).

To establish the sufficiency of the second part of (2.4), let us assume that the $\binom{n}{k}$ functions

$$\xi_{(i_1,\ldots,i_k)} : U \to R \qquad 1 < i_1 < \cdots < i_k \leq n$$

are smooth. To prove that ξ is smooth, let us consider an arbitrary k-tuple

$$\sigma = (V_1, \ldots, V_k)$$

of smooth vector fields $V_1, \ldots, V_k$. According to (2.2), there exist kn smooth functions

$$a_{ij} : U \to R \qquad (i = 1, 2, \ldots, k; j = 1, 2, \ldots, n)$$

such that

$$V_i(x) = \sum_{j=1}^{n} a_{ij}(x) \frac{\partial}{\partial \phi_j}$$

holds for every $i = 1, 2, \ldots, k$ and every $x \in U$. Because of (1.7) and (I,4.18), this implies that

$$\xi_\sigma(x) = \xi(x)[V_1(x), \ldots, V_k(x)]$$

can be expressed bilinearly in $\xi_{(i_1,\ldots,i_k)}(x)$ and $a_{ij}(x)$. This proves that ξ_σ is smooth for every k-tuple σ of smooth vector fields on U. Hence ξ is smooth. This completes the proof of (2.4).∎

Finally, for an arbitrary set of $\binom{n}{k}$ functions

$$\xi_{(i_1,\ldots,i_k)} : U \to R \qquad 1 \leq i_1 < \cdots < i_k \leq n$$

the function

$$\xi : U \to \Lambda^k(U)$$

defined by

$$\xi(x) = \Sigma \xi_{(i_1,\ldots,i_k)}(x)[d_x(\phi_{i_1}) \wedge \cdots \wedge d_x(\phi_{i_k})]$$

for every $x \in U$ is obviously a k-form on U. According to the second part of (2.4), ξ is smooth iff the given functions $\xi_{(i_1,\ldots,i_k)}$ are smooth.

EXERCISES

2A. Show that the smoothness of a k-form ξ on a smooth n-manifold X is a *local property;* that is, ξ is smooth iff there exists an open cover $\mathcal{C}$ of X such that, for every open set $U \in \mathcal{C}$, the restriction $\xi \mid U$ is a smooth k-form on the smooth n-manifold U.

2B. Let $f_1, \ldots, f_n$ be smooth real-valued functions defined on a neighborhood of a point x in a smooth n-manifold X. Prove that these functions (suitably restricted) are the coordinate functions of a chart iff

$$d_x(f_1) \wedge \cdots \wedge d_x(f_n) \neq 0.$$

2C. Prove that a smooth n-manifold X is orientable iff it has a smooth n-form that is nowhere zero on X.

2D. Prove that the tangent bundle $T(X)$ of any smooth n-manifold X is orientable.

2E. Investigate possible generalizations of the concepts and the results given in this section to the case in which the smooth manifold X is infinite-dimensional.

3. Exterior Differentiation

Throughout the present section, we are concerned with the algebra

$$\mathbf{C}^*(X) = \sum_{k=0}^{n} \mathbf{C}^k(X)$$

of all smooth differential forms on any given smooth n-manifold X.

First of all, let us consider the linear space $\mathbf{C}^0(X)$ of all smooth 0-forms on X. By definition, $\mathbf{C}^0(X)$ consists of all smooth functions

$$f : X \to R.$$

Let $f \in \mathbf{C}^0(X)$ be arbitrarily given. For every $x \in X$, its differential $d_x(f)$ is a cotangent vector in $T^*(X,x)$. Hence we may define a 1-form

$$d(f) : X \to \Lambda^1(X)$$

by taking

$$[d(f)](x) = d_x(f)$$

for every $x \in X$. This 1-form $d(f)$ will be called the *exterior differential* of the given smooth function f.

Lemma 3.1. *The exterior differential $d(f)$ of any smooth function $f : X \to R$ is smooth.*

PROOF. Let $\xi = d(f)$ and let V denote an arbitrary smooth vector field on X. It is sufficient to prove that the function

$$\xi_V : X \to R$$

defined by

$$\xi_V(x) = \xi(x)[V(x)]$$

for every $x \in X$ is smooth. For this purpose, we observe that

$$\xi_V(x) = d_x(f)[V(x)] = \langle V(x), d_x(f)\rangle = [V(x)](f) = f_V(x)$$

holds for every $x \in X$, which proves that $\xi_V = f_V$. Since f and V are smooth, it follows from the definition of the smoothness of vector fields that f_V is smooth. This implies the smoothness of $\xi = d(f)$ and completes the proof of (3.1).▌

Because of (3.1), the assignment $f \to d(f)$ defines a function

$$d : \mathbf{C}^0(X) \to \mathbf{C}^1(X)$$

called the *exterior differentiation* on $\mathbf{C}^0(X)$.

The following proposition is an immediate consequence of (I,4.12).

Proposition 3.2. *The exterior differentiation d on $\mathbf{C}^0(X)$ has the following properties:*

(a) *d is a linear transformation; that is, for arbitrary $f, g \in \mathbf{C}^0(X)$ and $a, b \in R$, we have*

$$d(af + bg) = a[d(f)] + b[d(g)].$$

(b) *d is a derivation; that is, for arbitrary $f, g \in \mathbf{C}^0(X)$, we have*

$$d(f \wedge g) = [d(f)] \wedge g + f \wedge [d(g)].$$

Since the exterior products $f \wedge \xi$ and $\xi \wedge f$ of a 0-form $f : X \to R$ and a differential form $\xi : X \to \Lambda(X)$ are defined by

$$(f \wedge \xi)(x) = f(x)\xi(x) = (\xi \wedge f)(x)$$

for every $x \in X$, they will be denoted by the simpler notation $f\xi$.

Naturally, one might inquire whether this exterior differentiation

$$d : \mathbf{C}^0(X) \to \mathbf{C}^1(X)$$

can be uniquely extended to a function

$$e : \mathbf{C}^*(X) \to \mathbf{C}^*(X)$$

with a few desirable properties. The answer turns out to be affirmative. We shall first define the concept of an exterior differentiation in the algebra $\mathbf{C}^*(X)$.

By an *exterior differentiation* in the algebra $\mathbf{C}^*(X)$ of all smooth differential forms on X, we mean a linear transformation

$$e : \mathbf{C}^*(X) \to \mathbf{C}^*(X)$$

of the linear space $\mathbf{C}^*(X)$ into itself that has the following four properties:

(ED1) e is an *extension* of the exterior differentiation $d : \mathbf{C}^0(X) \to \mathbf{C}^1(X)$ on $\mathbf{C}^0(X)$; that is, for every $f \in \mathbf{C}^0(X)$, we have

$$e(f) = d(f).$$

(ED2) e is *homogeneous of degree* 1; that is, for every $\xi \in \mathbf{C}^k(X)$, we have

$$e(\xi) \in \mathbf{C}^{k+1}(X).$$

(ED3) e is a *derivation;* that is, for arbitrary $\xi \in \mathbf{C}^k(X)$ and $\eta \in \mathbf{C}^*(X)$, we have

$$e(\xi \wedge \eta) = [e(\xi)] \wedge \eta + (-1)^k \xi \wedge [e(\eta)].$$

(ED4) e is an *idempotent;* that is, the composition $e \circ e$ of e with itself is the constant zero transformation of $\mathbf{C}^*(X)$ into itself.

Lemma 3.3. *Every exterior differentiation e in the algebra $\mathbf{C}^*(X)$ has the following additional property:*

(ED5) For every smooth differential form $\xi : X \to \Lambda(X)$ such that $\xi(x) = 0$ for every point x of an open subset W of X, we have

$$[e(\xi)](x) = 0$$

for every $x \in W$.

PROOF. Let w denote an arbitrary point of W. According to (I,3.10), there exists a smooth function

$$f : X \to R$$

such that $f(w) = 1$ and $f(x) = 0$ for every $x \in X \setminus W$. Then

$$(f\xi)(x) = f(x)\xi(x) = 0$$

for every $x \in X$. This implies that $f\xi$ is the zero element 0 of $\mathbf{C}^*(X)$. Since e is linear, it follows from (ED1) and (ED3) that

$$[d(f)] \wedge \xi + f \wedge [e(\xi)] = e(f\xi) = e(0) = 0.$$

Consequently, we obtain

$$[e(\xi)](w) = f(w) \cdot [e(\xi)](w) = -[d(f) \wedge \xi](w) = 0$$

because $\xi(w) = 0$. Since this is true for every $w \in W$, we have proved (3.3).∎

The following corollary is an obvious consequence of (3.3).

Corollary 3.4. *If e is an exterior differentiation in the algebra* $\mathbf{C}^*(X)$, *then, for any two smooth differential forms* ξ, η *on* X *such that* $\xi(x) = \eta(x)$ *holds for every point* x *of an open subset* W *of* X, *we have*

$$[e(\xi)](x) = [e(\eta)](x)$$

for every $x \in W$.

Now let us establish the following main theorem about exterior differentiation in the algebra $\mathbf{C}^*(X)$.

Theorem 3.5. *For an arbitrary smooth n-manifold* X, *there is a unique exterior differentiation in the algebra* $\mathbf{C}^*(X)$ *of all smooth differential forms on* X.

PROOF. The following proof consists of three parts: the uniqueness, the local existence, and the global existence.

Uniqueness. Assume that e is an arbitrary exterior differentiation in $\mathbf{C}^*(X)$. To prove the uniqueness of e, it suffices to show that, for any smooth k-form

$$\xi: X \to \Lambda^k(X)$$

on X and every point $x \in X$, the element

$$[e(\xi)](x) \in \Lambda^{k+1}[T(X,x)]$$

does not depend on the choice of e.

For this purpose, let us consider an arbitrary smooth chart

$$\phi : U \to R^n$$

on X with $x \in U$. Since $\xi \mid U$ is a smooth k-form on U, it follows from (2.4) that there exist $\binom{n}{k}$ uniquely determined smooth functions

$$\xi_{(i_1, \ldots, i_k)} : U \to R \qquad 1 \leq i_1 < \cdots < i_k \leq n$$

such that

$$\xi(u) = \Sigma \xi_{(i_1, \ldots, i_k)}(u)[d_u(\phi_{i_1}) \wedge \cdots \wedge d_u(\phi_{i_k})]$$

holds for every $u \in U$.

Choose an open neighborhood V of x in X such that its closure Cl (V) is compact and is contained in U. According to (I,3.10), there exists a smooth function

$$f : X \to R$$

such that $f(w) = 1$ for every $w \in$ Cl (V) and that its support $S(f)$ is contained in U.

Define $\binom{n}{k} + n$ smooth functions

$$\eta_{(i_1, \ldots, i_k)} : X \to R \qquad 1 \leq i_1 < \cdots < i_k \leq n$$

and

$$\psi_j : X \to R \qquad j = 1, 2, \ldots, n$$

by taking

$$\eta_{(i_1, \ldots, i_k)}(w) = \begin{cases} f(w)\xi_{(i_1, \ldots, i_k)}(w) & \text{(if } w \in U) \\ 0 & \text{(otherwise)} \end{cases}$$

$$\psi_j(w) = \begin{cases} f(w)\phi_j(w) & \text{(if } w \in \dot{U}) \\ 0 & \text{(otherwise)} \end{cases}$$

for every $w \in X$. Then we obtain a smooth k-form

$$\eta = \Sigma\eta_{(i_1, \ldots, i_k)} \wedge d(\psi_{i_1}) \wedge \cdots \wedge d(\psi_{i_k})$$

on X that does not depend on the choice of e. Because of the condition (ED1), we also have

$$\eta = \Sigma\eta_{(i_1, \ldots, i_k)} \wedge e(\psi_{i_1}) \wedge \cdots \wedge e(\psi_{i_k}).$$

By means of a finite induction on k, we can easily deduce from (ED3) and (ED4) that

$$e[e(\psi_{i_1}) \wedge \cdots \wedge e(\psi_{i_k})] = 0.$$

Consequently, it follows from the linearity of e, together with the conditions (ED1) and (ED3), that

$$\begin{aligned} e(\eta) &= \Sigma e[\eta_{(i_1, \ldots, i_k)}] \wedge e(\psi_{i_1}) \wedge \cdots \wedge e(\psi_{i_k}) \\ &= \Sigma d[\eta_{(i_1, \ldots, i_k)}] \wedge d(\psi_{i_1}) \wedge \cdots \wedge d(\psi_{i_k}). \end{aligned}$$

Therefore, the smooth $(k + 1)$-form $e(\eta)$ does not depend on the choice of e.

By the definitions of the smooth functions $\eta_{(i_1, \ldots, i_k)}$ and ψ_j, it is clear that

$$\begin{aligned} \eta(u) &= \Sigma\eta_{(i_1, \ldots, i_k)}(u)[d_u(\psi_{i_1}) \wedge \cdots \wedge d_u(\psi_{i_k})] \\ &= \Sigma\xi_{(i_1, \ldots, i_k)}(u)[d_u(\phi_{i_1}) \wedge \cdots \wedge d_u(\phi_{i_k})] = \xi(u) \end{aligned}$$

for every $u \in V$. Since $x \in V$, it follows from (3.4) that

$$[e(\xi)](x) = [e(\eta)](x).$$

Since $e(\eta)$ does not depend on the choice of e, this completes the uniqueness proof.

Local Existence. Consider any smooth chart

$$\phi : U \to R^n$$

on X. Then U is a smooth n-manifold. We shall construct an exterior differentiation

$$e_\phi : \mathbf{C}^*(U) \to \mathbf{C}^*(U)$$

as follows. Let $\xi : U \to \Lambda^k(U)$ denote an arbitrary smooth k-form on U.

By (2.4), there exist $\binom{n}{k}$ uniquely determined smooth functions

$$\xi_{(i_1, \ldots, i_k)} : U \to R \qquad 1 \leq i_1 < \cdots < i_k \leq n$$

such that

$$\xi = \Sigma \xi_{(i_1, \ldots, i_k)} \wedge d(\phi_{i_1}) \wedge \cdots \wedge d(\phi_{i_k}).$$

Let $e_\phi(\xi)$ be the smooth $(k+1)$-form on U defined by

$$e_\phi(\xi) = \Sigma d[\xi_{(i_1, \ldots, i_k)}] \wedge d(\phi_{i_1}) \wedge \cdots \wedge d(\phi_{i_k}).$$

By the aid of linearity, the assignment $\xi \to e_\phi(\xi)$ determines a linear transformation

$$e_\phi : \mathbf{C}^*(U) \to \mathbf{C}^*(U).$$

To verify that e_ϕ is an exterior differentiation in $\mathbf{C}^*(U)$, we have to verify the four properties (ED1)–(ED4). Since (ED1) and (ED2) are obvious, it remains to establish (ED3) and (ED4).

To verify (ED3), let $\xi \in \mathbf{C}^k(U)$ and $\eta \in \mathbf{C}^l(U)$ be arbitrarily given. Because of the linearity of e_ϕ and the bilinearity of $\wedge$, we may assume that ξ and η are of the form

$$\xi = f \wedge d(\phi_{i_1}) \wedge \cdots \wedge d(\phi_{i_k}) \qquad \eta = g \wedge d(\phi_{j_1}) \wedge \cdots \wedge d(\phi_{j_l})$$

where $f, g : U \to R$ are two smooth functions and

$$1 \leq i_1 < \cdots < i_k \leq n \qquad 1 \leq j_1 < \cdots < j_l \leq n.$$

Then we have

$$\xi \wedge \eta = (fg) \wedge d(\phi_{i_1}) \wedge \cdots \wedge d(\phi_{i_k}) \wedge d(\phi_{j_1}) \wedge \cdots \wedge d(\phi_{j_l}).$$

By (b) of (3.2) and (c) of (2.3), we deduce that

$$\begin{aligned}
e_\phi(\xi \wedge \eta) &= d(fg) \wedge d(\phi_{i_1}) \wedge \cdots \wedge d(\phi_{i_k}) \wedge d(\phi_{j_1}) \wedge \cdots \wedge d(\phi_{j_l}) \\
&= d(f) \wedge g \wedge d(\phi_{i_1}) \wedge \cdots \wedge d(\phi_{i_k}) \wedge d(\phi_{j_1}) \wedge \cdots \wedge d(\phi_{j_l}) \\
&\quad + f \wedge d(g) \wedge d(\phi_{i_1}) \wedge \cdots \wedge d(\phi_{i_k}) \wedge d(\phi_{j_1}) \wedge \cdots \wedge d(\phi_{j_l}) \\
&= [d(f) \wedge d(\phi_{i_1}) \wedge \cdots \wedge d(\phi_{i_k})] \wedge [g \wedge d(\phi_{j_1}) \wedge \cdots \wedge d(\phi_{j_l})] \\
&\quad + (-1)^k [f \wedge d(\phi_{i_1}) \wedge \cdots \wedge d(\phi_{i_k})] \wedge [d(g) \wedge d(\phi_{j_1}) \wedge \cdots \wedge d(\phi_{j_l})] \\
&= e_\phi(\xi) \wedge \eta + (-1)^k \xi \wedge e_\phi(\eta).
\end{aligned}$$

This establishes (ED3).

To verify (ED4), let $\xi \in \mathbf{C}^k(U)$ be arbitrarily given. As before, we may assume that ξ is of the form

$$\xi = f \wedge d(\phi_{i_1}) \wedge \cdots \wedge d(\phi_{i_k})$$

where $f : U \to R$ is a smooth function and

$$1 \leq i_1 < \cdots < i_k \leq n.$$

Then we have

$$e_\phi(\xi) = d(f) \wedge d(\phi_{i_1}) \wedge \cdot \cdot \cdot \wedge d(\phi_{i_k}).$$

Consider the smooth function

$$g : \phi(U) \to R$$

defined on the open subset $\phi(U)$ of R^n by

$$g(t) = f[\phi^{-1}(t)]$$

for every $t \in \phi(U)$ and its partial derivatives

$$g_j = \frac{\partial g}{\partial t_j} : \phi(U) \to R \qquad (j = 1, 2, \ldots, n).$$

Then $g_1, \ldots, g_n$ are smooth functions and so are the compositions

$$f_j = g_j \circ \phi : U \to R \qquad (j = 1, 2, \ldots, n).$$

By (I,4.13), we have

$$d(f) = \sum_{j=1}^{n} f_j \wedge d(\phi_j).$$

Consequently,

$$e_\phi(\xi) = \sum_{j=1}^{n} f_j \wedge d(\phi_j) \wedge d(\phi_{i_1}) \wedge \cdot \cdot \cdot \wedge d(\phi_{i_k}).$$

Applying e_ϕ once more, we obtain

$$e_\phi[e_\phi(\xi)] = \sum_{j=1}^{n} e_\phi[f_j \wedge d(\phi_j) \wedge d(\phi_{i_1}) \wedge \cdot \cdot \cdot \wedge d(\phi_{i_k})]$$

$$= \sum_{j=1}^{n} \sum_{h=1}^{n} f_{jh} \wedge d(\phi_h) \wedge d(\phi_j) \wedge d(\phi_{i_1}) \wedge \cdot \cdot \cdot \wedge d(\phi_{i_k})]$$

where f_{jh} stands for the smooth function

$$f_{jh} = \frac{\partial^2 g}{\partial t_h \, \partial t_j} \circ \phi : U \to R.$$

Because

$$f_{hj} \wedge d(\phi_j) \wedge d(\phi_h) = -f_{jh} \wedge d(\phi_h) \wedge d(\phi_j)$$

for all $h, j = 1, 2, \ldots, n$, we obtain

$$e_\phi[e_\phi(\xi)] = 0.$$

This establishes (ED4) and completes the local existence proof.

Global Existence. Consider a smooth atlas σ of X. For each chart $\phi : U \to R^n$ in σ, we have constructed an exterior differentiation

$$e_\phi : \mathbf{C}^*(U) \to \mathbf{C}^*(U).$$

For any charts $\phi : U \to R^n$ and $\psi : V \to R^n$ in σ with $U \cap V \neq \square$, the restrictions $\phi \mid U \cap V$ and $\psi \mid U \cap V$ are also charts in σ. By the uniqueness of exterior differentiation in $\mathbf{C}^*(U \cap V)$, we have

$$e_{\phi|U\cap V} = e_{\psi|U\cap V}.$$

Now let $\xi : X \to \Lambda^k(X)$ denote an arbitrary smooth k-form on X. Define a $(k + 1)$-form

$$e(\xi) : X \to \Lambda^{k+1}(X)$$

as follows. Let x be any point of X. Then there exists a chart $\phi : U \to R^n$ in σ with $x \in U$. The restriction

$$\xi \mid U : U \to \Lambda^k(U)$$

is a smooth k-form on U. We shall show that the element

$$[e(\xi)](x) = [e_\phi(\xi \mid U)](x) \in \Lambda^{k+1}[T(X,x)]$$

does not depend on the choice of the chart $\phi : U \to R^n$. For this purpose, let $\psi : V \to R^n$ be another chart in σ with $x \in V$. Then, by the construction in the local existence proof, we have

$$\begin{aligned} [e_\phi(\xi \mid U)](x) &= [e_{\phi|U\cap V}(\xi \mid U \cap V)](x) \\ [e_\psi(\xi \mid V)](x) &= [e_{\psi|U\cap V}(\xi \mid U \cap V)](x). \end{aligned}$$

Since $e_{\phi|U\cap V} = e_{\psi|U\cap V}$, these imply

$$[e_\phi(\xi \mid U)](x) = [e_\psi(\xi \mid V)](x).$$

This proves that $[e(\xi)](x)$ does not depend on the choice of the chart $\phi : U \to R^n$. Hence the assignment $x \to [e(\xi)](x)$ defines a $(k + 1)$-form

$$e(\xi) : X \to \Lambda^{k+1}(X).$$

Since smoothness of differential forms is a local property according to Exercise 2A, the relation

$$e(\xi) \mid U = e_\phi(\xi \mid U)$$

for every chart $\phi : U \to R^n$ implies that $e(\xi)$ is a smooth $(k + 1)$-form on X. Hence the assignment $\xi \to e(\xi)$ determines a linear transformation

$$e : \mathbf{C}^*(X) \to \mathbf{C}^*(X).$$

The conditions (ED1)–(ED4) for e are obvious consequences of the same conditions for $e_\phi, \phi \in \sigma$. Hence e is an exterior differentiation in $\mathbf{C}^*(X)$. This completes the proof of (3.5).∎

We have used the notation

$$e : \mathbf{C}^*(X) \to \mathbf{C}^*(X)$$

for the exterior differentiation in $\mathbf{C}^*(X)$ to distinguish it from the ordinary exterior differentiation

$$d : \mathbf{C}^0(X) \to \mathbf{C}^1(X).$$

Having established the uniqueness and the existence of exterior differentiation in $\mathbf{C}^*(X)$, we no longer have to preserve the distinction between d and e. Hereafter, we shall use the usual notation

$$d : \mathbf{C}^*(X) \to \mathbf{C}^*(X)$$

for the uniquely determined *exterior differentiation* in the algebra $\mathbf{C}^*(X)$ of all smooth differential forms on X.

EXERCISES

3A. Verify $d\xi = 0$ for every constant function $\xi : X \to R$ in $\mathbf{C}^0(X)$.

3B. Let $\phi : U \to R^n$ be any smooth chart on a smooth n-manifold X. Show that the n coordinate functions

$$\phi_1, \ldots, \phi_n : U \to R$$

are 0-forms on U. Prove that, for an arbitrary k-form

$$\xi : U \to \Lambda^k(U)$$

on U, there exists $\binom{n}{k}$ uniquely determined 0-forms

$$\xi_{(i_1, \ldots, i_k)} \qquad 1 \leq i_1 < \cdots < i_k \leq n$$

on U such that

$$\xi = \Sigma \xi_{(i_1, \ldots, i_k)}[d\phi_{i_1} \wedge \cdots \wedge d\phi_{i_k}].$$

Finally, prove that, for an arbitrary set of $\binom{n}{k}$ 0-forms

$$\xi_{(i_1, \ldots, i_k)} \qquad 1 \leq i_1 < \cdots < i_k \leq n$$

on U,

$$\Sigma \xi_{(i_1, \ldots, i_k)}[d\phi_{i_1} \wedge \cdots \wedge d\phi_{i_k}]$$

is a k-form on U. In particular, the 1-forms on U are the expressions

$$\xi_1 \, d\phi_1 + \cdots + \xi_n \, d\phi_n$$

where $\xi_1, \ldots, \xi_n$ are arbitrary 0-forms on U.

3C. The coordinate functions of the Euclidean 3-space R^3 are usually denoted by x, y, and z. Deduce from Exercise 3B that the k-forms, $(k = 1, 2, 3)$, on R^3 are the expressions

$$\alpha dx + \beta dy + \gamma dz$$
$$\lambda(dy \wedge dz) + \mu(dz \wedge dx) + \nu(dx \wedge dy)$$
$$\theta(dx \wedge dy \wedge dz)$$

where α, β, γ, λ, μ, ν, and θ are arbitrary 0-forms on R^3—that is, arbitrary smooth real-valued functions on R^3. Establish the following statements:

(a) If ξ is any 0-form on R^3, then

$$d\xi = \frac{\partial \xi}{\partial x} dx + \frac{\partial \xi}{\partial y} dy + \frac{\partial \xi}{\partial z} dz.$$

(b) If $\xi = \alpha dx + \beta dy + \gamma dz$, then

$$d\xi = \left(\frac{\partial \gamma}{\partial y} - \frac{\partial \beta}{\partial z}\right)(dy \wedge dz) + \left(\frac{\partial \alpha}{\partial z} - \frac{\partial \gamma}{\partial x}\right)(dz \wedge dx) + \left(\frac{\partial \beta}{\partial x} - \frac{\partial \alpha}{\partial y}\right)(dx \wedge dy).$$

(c) If $\xi = \lambda(dy \wedge dz) + \mu(dz \wedge dx) + \nu(dx \wedge dy)$, then

$$d\xi = \left(\frac{\partial \lambda}{\partial x} + \frac{\partial \mu}{\partial y} + \frac{\partial \nu}{\partial z}\right)(dx \wedge dy \wedge dz).$$

3D. By a *smooth vector field* on R^3, we mean a smooth map

$$v : R^3 \to R^3.$$

Composing v with the coordinate functions of R^3, we obtain three smooth real-valued functions

$$\xi_v, \eta_v, \zeta_v : R^3 \to R$$

known as the components of the vector field v. For every point p of R^3, $\xi_v(p)$, $\eta_v(p)$, and $\zeta_v(p)$ are the coordinates of the point $v(p)$ in R^3. Establish the following statements:

(a) For every smooth function $\phi : R^3 \to R$, the three partial derivatives

$$\frac{\partial \phi}{\partial x}, \frac{\partial \phi}{\partial y}, \frac{\partial \phi}{\partial z} : R^3 \to R$$

are the components of a smooth vector field, which is known as the *gradient* of ϕ and is denoted by grad (ϕ).

(b) For every smooth vector field v on R^3, the three functions

$$\frac{\partial \zeta_v}{\partial y} - \frac{\partial \eta_v}{\partial z}, \quad \frac{\partial \xi_v}{\partial z} - \frac{\partial \zeta_v}{\partial x}, \quad \frac{\partial \eta_v}{\partial x} - \frac{\partial \xi_v}{\partial y} \quad : R^3 \to R$$

are the components of a smooth vector field, which is known as the *curl* of v and is denoted by curl (v).

(c) For every smooth vector field v on R^3, the function

$$\frac{\partial \xi_v}{\partial x} + \frac{\partial \eta_v}{\partial y} + \frac{\partial \zeta_v}{\partial z} : R^3 \to R$$

is smooth. This smooth function is known as the *divergent* of v and is denoted by div (v).

3E. Prove that the set V of all vector fields on R^3 forms a linear space over R. Verify that the two functions

$$\Phi : V \to \mathbf{C}^1(R^3) \qquad \Psi : V \to \mathbf{C}^2(R^3)$$

defined by

$$\Phi(v) = \xi_v\, dx + \eta_v\, dy + \zeta_v\, dz$$

$$\Psi(v) = \xi_v(dy \wedge dz) + \eta_v(dz \wedge dx) + \zeta_v(dx \wedge dy)$$

for every vector field v on R^3 are isomorphisms of linear spaces over R. Deduce the following assertions from Exercises 3C and 3D:

(a) For every 0-form ω on R^3, we have

$$d\omega = \Phi\,[\text{grad}\,(\omega)].$$

(b) For every smooth vector field v on R^3, we have

$$d\Phi(v) = \Psi[\text{curl}\,(v)].$$

(c) For every smooth vector field v on R^3, we have

$$d\Psi(v) = \text{div}\,(v)\,[dx \wedge dy \wedge dz].$$

3F. By means of Exercise 3E and the property $d \circ d = 0$ of exterior differentiation, prove the following two statements:

(a) For every smooth function $f : R^3 \to R$, we have

$$\text{curl}\,[\text{grad}\,(f)] = 0.$$

(b) For every smooth vector field $v : R^3 \to R^3$, we have

$$\text{div}\,[\text{curl}\,(v)] = 0.$$

3G. Investigate possible generalizations of the concepts and the results in this section to the case where the smooth manifold X is infinite-dimensional.

4. De Rham Cohomology Groups

Let X denote an arbitrarily given smooth n-manifold. In §2, we constructed the algebra

$$\mathbf{C}^*(X) = \sum_{k=0}^{n} \mathbf{C}^k(X)$$

of all smooth differential forms on X. In §3, we established the existence of a uniquely determined exterior differentiation

$$d : \mathbf{C}^*(X) \to \mathbf{C}^*(X).$$

Since d is homogeneous of degree 1 according to (ED2) in §3, it can be developed into an infinite sequence

$$\cdots \to \mathbf{C}^{k-1}(X) \xrightarrow{d} \mathbf{C}^k(X) \xrightarrow{d} \mathbf{C}^{k+1}(X) \to \cdots$$

of linear transformations with the understanding that

$$\mathbf{C}^k(X) = 0$$

if $k < 0$ or $k > n$. Because of

$$d \circ d = 0$$

according to (ED4) in §3, this infinite sequence of linear transformations is said to be *semi-exact* or a *cochain complex*.

For every integer k, let $\mathbf{Z}^k(X)$ denote the kernel of the linear transformation

$$d : \mathbf{C}^k(X) \to \mathbf{C}^{k+1}(X)$$

that is,

$$\mathbf{Z}^k(X) = \{\xi \in \mathbf{C}^k(X) \mid d(\xi) = 0\}.$$

Then $\mathbf{Z}^k(X)$ is a linear subspace of $\mathbf{C}^k(X)$. The members of $\mathbf{Z}^k(X)$ are said to be the *closed smooth k-forms* on X. The following lemma is obvious.

Lemma 4.1. *For every smooth n-manifold X, we have*

$$\mathbf{Z}^n(X) = \mathbf{C}^n(X).$$

On the other hand, let $\mathbf{B}^k(X)$ denote the image of the linear transformation

$$d : \mathbf{C}^{k-1}(X) \to \mathbf{C}^k(X)$$

that is,

$$\mathbf{B}^k(X) = \{d(\xi) \in \mathbf{C}^k(X) \mid \xi \in \mathbf{C}^{k-1}(X)\}.$$

Then $\mathbf{B}^k(X)$ is also a linear subspace of $\mathbf{C}^k(X)$. The members of $\mathbf{B}^k(X)$ are said to be the *exact smooth k-forms* on X. The following lemma is obvious.

Lemma 4.2. *For every smooth n-manifold X, we have*

$$\mathbf{B}^0(X) = 0.$$

Now let us establish the following important lemma, which is rather obvious.

Lemma 4.3. *Every exact smooth k-form on X is closed.*

PROOF. Let η denote any exact smooth k-form on X. Then there exists a smooth $(k-1)$-form ξ on X with $\eta = d(\xi)$. Since $d \circ d = 0$, we have

$$d(\eta) = d[d(\xi)] = (d \circ d)(\xi) = 0.$$

This implies that η is closed.▌

The following corollary is a direct consequence of (4.3).

Corollary 4.4. *For every smooth n-manifold X and any integer k,* $\mathbf{B}^k(X)$ *is a linear subspace of* $\mathbf{Z}^k(X)$; *in symbols,*

$$\mathbf{B}^k(X) \subset \mathbf{Z}^k(X).$$

On the basis of (4.4), we obtain a quotient linear space

$$\mathbf{H}^k(X) = \mathbf{Z}^k(X)/\mathbf{B}^k(X)$$

over R for every integer k. This linear space $\mathbf{H}^k(X)$ will be referred to as the *k-dimensional de Rham cohomology group* of the given smooth n-manifold X.

The following proposition is a direct consequence of (4.1) and (4.2).

Proposition 4.5. *For every smooth n-manifold X, we have*

$$\mathbf{H}^k(X) = \begin{cases} \mathbf{Z}^0(X) & (\textit{if } k = 0) \\ \mathbf{C}^n(X)/\mathbf{B}^n(X) & (\textit{if } k = n) \\ 0 & (\textit{if } k < 0 \text{ or } k > n). \end{cases}$$

Now let $\mathbf{Z}^*(X)$ and $\mathbf{B}^*(X)$ denote the kernel and the image of the linear transformation

$$d : \mathbf{C}^*(X) \to \mathbf{C}^*(X)$$

respectively. Then obviously we have the following direct sum decompositions:

$$\mathbf{Z}^*(X) = \sum_{k=0}^{n} \mathbf{Z}^k(X)$$

$$\mathbf{B}^*(X) = \sum_{k=0}^{n} \mathbf{B}^k(X).$$

Thus $\mathbf{B}^*(X)$ is a linear subspace of $\mathbf{Z}^*(X)$. The quotient linear space

$$\mathbf{H}^*(X) = \mathbf{Z}^*(X)/\mathbf{B}^*(X)$$

can be easily identified with the direct sum

$$\mathbf{H}^*(X) = \sum_{k=0}^{n} \mathbf{H}^k(X).$$

Define a bilinear multiplication $\wedge$ in the linear space $\mathbf{H}^*(X)$ as follows: Let $\alpha \in \mathbf{H}^k(X)$ and $\beta \in \mathbf{H}^l(X)$ be arbitrarily given. Choose $\xi \in \mathbf{Z}^k(X)$ and $\eta \in \mathbf{Z}^l(X)$, which represent the cohomology classes α and β respectively.

Lemma 4.6. *The exterior product $\xi \wedge \eta$ is a closed smooth $(k + l)$-form on X.*

PROOF. According to (b) of (2.3), $\xi \wedge \eta$ is a smooth $(k + l)$-form on X. It remains to show that $\xi \wedge \eta$ is closed. For this purpose, we observe that (ED3) in §3 implies that

$$d(\xi \wedge \eta) = d(\xi) \wedge \eta + (-1)^k \xi \wedge d(\eta) = 0$$

since $d(\xi) = 0$ and $d(\eta) = 0$. This proves that $\xi \wedge \eta$ is closed.∎

Because of (4.6), $\xi \wedge \eta$ represents a cohomology class

$$[\xi \wedge \eta] \in \mathbf{H}^{k+l}(X).$$

Lemma 4.7. *The cohomology class $[\xi \wedge \eta]$ in $\mathbf{H}^{k+l}(X)$ does not depend on the choice of $\xi \in \alpha$ and $\eta \in \beta$.*

PROOF. For any other choice $\xi' \in \alpha$ and $\eta' \in \beta$,

$$\xi' - \xi \in \mathbf{B}^k(X) \qquad \eta' - \eta \in \mathbf{B}^l(X).$$

Therefore, there exist $\lambda \in \mathbf{C}^{k-1}(X)$ and $\mu \in \mathbf{C}^{l-1}(X)$ with

$$d\lambda = \xi' - \xi \qquad d\mu = \eta' - \eta.$$

Since ξ and η' are closed,

$$d(\xi \wedge \mu) = (-1)^k \xi \wedge d\mu \qquad d(\lambda \wedge \eta') = d\lambda \wedge \eta'.$$

Consequently, we obtain

$$\begin{aligned}\xi' \wedge \eta' &= \xi \wedge \eta' + d\lambda \wedge \eta' \\ &= \xi \wedge \eta + \xi \wedge d\mu + d\lambda \wedge \eta' \\ &= \xi \wedge \eta + d((-1)^k \xi \wedge \mu + \lambda \wedge \eta').\end{aligned}$$

This implies that $\xi \wedge \eta$ and $\xi' \wedge \eta'$ represent the same cohomology class in $\mathbf{H}^{k+l}(X)$ and completes the proof of (4.7).▌

Because of (4.6) and (4.7), we may define a product $\alpha \wedge \beta$ of α and β by taking

$$\alpha \wedge \beta = [\xi \wedge \eta] \in \mathbf{H}^{k+l}(X)$$

for any $\xi \in \alpha$ and $\eta \in \beta$. By means of bilinearity, the assignment $(\alpha,\beta) \to \alpha \wedge \beta$ determines a bilinear operation

$$\wedge : \mathbf{H}^*(X) \times \mathbf{H}^*(X) \to \mathbf{H}^*(X).$$

This bilinear operation in $\mathbf{H}^*(X)$ will be referred to as the *exterior multiplication*. The following theorem is obvious.

Theorem 4.8. *The exterior multiplication* $\wedge$ *in* $\mathbf{H}^*(X)$ *makes* $\mathbf{H}^*(X)$ *an associative algebra over* R *with the following properties:*

(a) *The element* $1 \in \mathbf{H}^0(X)$ *represented by the constant function* 1 *on* X *is a two-sided multiplicative unit, that is,*

$$1 \wedge \alpha = \alpha = \alpha \wedge 1$$

holds for every element α *of* $\mathbf{H}^*(X)$.

(b) *For arbitrary* $\alpha \in \mathbf{H}^k(X)$ *and* $\beta \in \mathbf{H}^l(X)$, *we have*

$$\beta \wedge \alpha = (-1)^{kl} \alpha \wedge \beta \in \mathbf{H}^{k+l}(X).$$

This associative algebra $\mathbf{H}^*(X)$ over R will be called the *de Rham cohomology algebra* of the given smooth n-manifold X.

EXERCISES

4A. For any connected smooth n-manifold X, prove that

$$\mathbf{H}^0(X) \approx R.$$

4B. For any two disjoint smooth n-manifolds X and Y, let $X + Y$ denote their *topological sum* as defined in Example (d) of (I, §2). Prove that

$$\mathbf{H}^k(X + Y) \approx \mathbf{H}^k(X) \oplus \mathbf{H}^k(Y)$$

holds for every integer k. Here $\oplus$ stands for direct sum.

4C. By means of your knowledge of calculus, prove that

$$\mathbf{H}^k(R^n) = 0$$

holds for every $n = 1, 2, 3$ and every $k > 0$.

4D. Investigate possible generalizations of the concepts and the results given in this section to the case where the smooth manifold X is infinite-dimensional.

5. Induced Transformations

Let us consider an arbitrary smooth map

$$f : X \to Y$$

from a smooth m-manifold X into a smooth n-manifold Y. For every $x \in X$, we have defined in (I, §5), a linear transformation

$$f_* = T(f,x) : T(X,x) \to T(Y,y)$$

with $y = f(x)$. Now we shall define a linear transformation

$$f^\# = \mathbf{F}^k(f) : \mathbf{F}^k(Y) \to \mathbf{F}^k(X)$$

of the linear spaces of k-forms as follows. In case $k = 0$, $f^\#$ is defined by taking

$$f^\#(\xi) = \xi \circ f \in \mathbf{F}^0(X)$$

for every function $\xi : Y \to R$ in $\mathbf{F}^0(Y)$. In case $k > 0$, $f^\#$ is defined by taking

$$\{[f^\#(\xi)](x)\}(v_1, \ldots, v_k) = [\xi(y)](f_*v_1, \ldots, f_*v_k)$$

for every $\xi \in \mathbf{F}^k(Y)$ and arbitrary $v_1, \ldots, v_k$ in $T(X,x)$.

Passing to the direct sums, we obtain a linear transformation

$$f^\# = \mathbf{F}^*(f) : \mathbf{F}^*(Y) \to \mathbf{F}^*(X).$$

For every $\xi \in \mathbf{F}^k(Y)$ and every $\eta \in \mathbf{F}^l(Y)$, we have

$$\begin{aligned}\{[f^\#(\xi \wedge \eta)](x)\}(v_1, \ldots, v_{k+l}) &= [(\xi \wedge \eta)(y)](f_*v_1, \ldots, f_*v_{k+l}) \\ &= \{[f^\#(\xi) \wedge f^\#(\eta)](x)\}(v_1, \ldots, v_{k+l})\end{aligned}$$

for every $x \in X$ and arbitrary $v_1, \ldots, v_{k+l}$ in $T(X,x)$. This implies

$$f^\#(\xi \wedge \eta) = f^\#(\xi) \wedge f^\#(\eta)$$

and hence $f^\# = \mathbf{F}^*(f)$ is a homomorphism of the algebra $\mathbf{F}^*(Y)$ into the algebra $\mathbf{F}^*(X)$. We have established the following lemma:

Lemma 5.1. *The transformation*

$$\mathbf{F}^*(f) : \mathbf{F}^*(Y) \to \mathbf{F}^*(X)$$

is a homomorphism of the algebra $\mathbf{F}^*(Y)$ *into the algebra* $\mathbf{F}^*(X)$ *with the following two properties:*

(a) $\mathbf{F}^*(f)$ *is homogeneous of degree* 0; *that is, it sends* $\mathbf{F}^k(Y)$ *into* $\mathbf{F}^k(X)$ *for every* k.

(b) $\mathbf{F}^*(f)$ *preserves multiplicative unit; that is, it sends the unit of* $\mathbf{F}^*(Y)$ *to the unit of* $\mathbf{F}^*(X)$.

Next let us establish the following lemma:

Lemma 5.2. *The homomorphism* $\mathbf{F}^*(f)$ *preserves smoothness; that is, it carries the subalgebra* $\mathbf{C}^*(Y)$ *of* $\mathbf{F}^*(Y)$ *into the subalgebra* $\mathbf{C}^*(X)$ *of* $\mathbf{F}^*(X)$.

PROOF. Let ξ denote an arbitrary smooth k-form on Y. It suffices to verify that $f^{\#}(\xi)$ is a smooth k-form on X.

First, let us consider the case $k = 0$. Then $\xi : Y \to R$ is a smooth function. By definition, $f^{\#}(\xi)$ is the composition

$$f^{\#}(\xi) = \xi \circ f : X \to R$$

and hence is a smooth function on X. This verifies the case $k = 0$.

Next, let us consider the case $\xi = d(\eta)$ with $\eta \in \mathbf{C}^0(Y)$. To prove that the 1-form $\xi^* = f^{\#}(\xi)$ on X is smooth, we have to show that the function

$$\xi^*_V : X \to R$$

defined for an arbitrarily given smooth tangent field V on X by

$$\xi^*_V(x) = \xi^*(x)[V(x)]$$

for every $x \in X$, is smooth. By definition of $\xi^* = f^{\#}(\xi)$,

$$\xi^*(x)[V(x)] = \xi(y)\{f_*[V(x)]\}$$

with $y = f(x)$. Since $\xi = d(\eta)$, it follows from (I,4.18) that

$$\xi(y)\{f_*[V(x)]\} = d_y(\eta)\{f_*[V(x)]\} = \{f_*[V(x)]\}[d_y(\eta)].$$

Finally, it follows from the definition of f_* in (5,§5) that

$$\{f_*[V(x)]\}[d_y(\eta)] = [V(x)](\eta \circ f).$$

Combining these equalities, we obtain

$$\xi^*_V(x) = [V(x)](\eta \circ f)$$

for every $x \in X$. Since V is a smooth vector field on X and $\eta \circ f : X \to R$ is a smooth function, it follows that ξ^*_V is smooth. This implies that $\xi^* = f^{\#}(\xi)$ is smooth for this case.

Now let us consider the general case. Let ξ denote an arbitrary smooth k-form on Y. To prove the smoothness of the k-form $\xi^* = f^{\#}(\xi)$, let x denote an arbitrary point of X and let $y = f(x)$. Select a smooth chart

$$\phi : U \to R^n$$

on Y with $y \in U$. According to (2.4), there exist $\binom{n}{k}$ uniquely smooth functions

$$\xi_{(i_1, \ldots, i_k)} : U \to R \qquad 1 \leq i_1 < \cdots < i_k \leq n$$

such that

$$\xi(u) = \Sigma \xi_{(i_1, \ldots, i_k)}(u)[d_u(\phi_{i_1}) \wedge \cdots \wedge d_u(\phi_{i_k})]$$

holds for every point $u \in U$.

Choose an open neighborhood V of y in Y such that its closure Cl (V) is compact and is contained in U. According to (I,3.10), there exists a smooth function

$$\lambda : Y \to R$$

such that $\lambda(w) = 1$ for every $w \in$ Cl (V) and that its support $S(\lambda)$ is contained in U.

Define $\binom{n}{k} + n$ smooth functions

$$\eta_{(i_1, \ldots, i_k)} : Y \to R \qquad 1 \leq i_1 < \cdots < i_k \leq n$$

and

$$\psi_j : Y \to R \qquad j = 1, 2, \ldots, n$$

by taking

$$\eta_{(i_1, \ldots, i_k)}(w) = \begin{cases} \lambda(w)\xi_{(i_1, \ldots, i_k)}(w) & \text{(if } w \in U) \\ 0 & \text{(otherwise)} \end{cases}$$

$$\psi_j(w) = \begin{cases} \lambda(w)\phi_j(w) & \text{(if } w \in U) \\ 0 & \text{(otherwise)} \end{cases}$$

for every $w \in Y$. Then we obtain a smooth k-form

$$\eta = \Sigma \eta_{(i_1, \ldots, i_k)} \wedge d(\psi_{i_1}) \wedge \cdots \wedge d(\psi_{i_k})$$

on Y.

Since $f^{\#}$ is a homomorphism of the algebra $\mathbf{F}^*(Y)$ into the algebra $\mathbf{F}^*(X)$, we have

$$f^{\#}(\eta) = \Sigma f^{\#}[\eta_{(i_1, \ldots, i_k)}] \wedge f^{\#}[d(\psi_{i_1})] \wedge \cdots \wedge f^{\#}[d(\psi_{i_k})].$$

By the two special cases which have been proved, the $\binom{n}{k} + n$ differential forms

$$f^{\#}[\eta_{(i_1, \ldots, i_k)}] \qquad f^{\#}[d(\psi_j)]$$

are smooth. Since $\mathbf{C}^*(X)$ is a subalgebra of $\mathbf{F}^*(X)$, it follows that $f^{\#}(\eta)$ is smooth. Let

$$W = f^{-1}(V).$$

Then W is an open neighborhood of x in X. Since

$$f^{\#}(\xi) \mid W = f^{\#}(\eta) \mid W$$

$f^{\#}(\xi)$ is smooth in W. Since smoothness is a local property and x is any point of X, this implies that $f^{\#}(\xi)$ is smooth. This completes the proof of (5.2).∎

On the basis of (5.2), $\mathbf{F}^*(f)$ defines a homomorphism

$$\mathbf{C}^*(f) : \mathbf{C}^*(Y) \to \mathbf{C}^*(X)$$

of the algebra $\mathbf{C}^*(Y)$ of all smooth differential forms on Y into the algebra $\mathbf{C}^*(X)$ of all smooth differential forms on X. By (a) and (b), $\mathbf{C}^*(f)$ is homogeneous of degree 0 and preserves multiplicative unit. For every integer k, $\mathbf{C}^*(f)$ defines a linear transformation

$$\mathbf{C}^k(f) : \mathbf{C}^k(Y) \to \mathbf{C}^k(X).$$

When there is no danger of ambiguity, all of these linear transformations will be denoted by the same symbol $f^{\#}$.

Lemma 5.3. *For an arbitrary smooth map $f : X \to Y$ and any integer k, the rectangle*

$$\begin{array}{ccc} \mathbf{C}^k(Y) & \xrightarrow{d} & \mathbf{C}^{k+1}(Y) \\ \downarrow f^{\#} & & \downarrow f^{\#} \\ \mathbf{C}^k(X) & \xrightarrow{d} & \mathbf{C}^{k+1}(X) \end{array}$$

is commutative; that is, we have

$$d \circ f^{\#} = f^{\#} \circ d.$$

PROOF. Let ξ denote an arbitrary smooth k-form on Y. It suffices to verify that

$$(d \circ f^{\#})(\xi) = (f^{\#} \circ d)(\xi).$$

First, let us consider the case $k = 0$. Then $\xi : Y \to R$ is a smooth function. Let $x \in X$ and $v \in T(X,x)$ be arbitrarily given. Write $y = f(x)$. Then we have

$$\begin{aligned} \{[(d \circ f^{\#})(\xi)](x)\}(v) &= \{[d(\xi \circ f)](x)\}(v) \\ &= [d_x(\xi \circ f)](v) \\ &= v(\xi \circ f) \\ \{[(f^{\#} \circ d)(\xi)](x)\}(v) &= \{[f^{\#}(d\xi)](x)\}(v) \\ &= [(d\xi)(y)]f_*(v) \\ &= [f_*(v)]d_y(\xi) \\ &= v(\xi \circ f). \end{aligned}$$

Since these hold for every $x \in X$ and every $v \in T(X,x)$, we have

$$(d \circ f^{\#})(\xi) = (f^{\#} \circ d)(\xi).$$

Next let us consider the case $\xi = d(\eta)$ with $\eta \in \mathbf{C}^0(Y)$. Then

$$(f^{\#} \circ d)(\xi) = f^{\#}[(d \circ d)(\eta)] = f^{\#}(0) = 0.$$

On the other hand,

$$(d \circ f^{\#})(\xi) = d[(f^{\#} \circ d)(\eta)].$$

Since $\eta \in \mathbf{C}^0(Y)$, we have

$$(f^{\#} \circ d)(\eta) = (d \circ f^{\#})(\eta).$$

According to the preceding special case, we obtain

$$(d \circ f^{\#})(\xi) = (d \circ d \circ f^{\#})(\eta) = 0 = (f^{\#} \circ d)(\xi)$$

because $d \circ d = 0$.

Now let us consider the general case. Then ξ is an arbitrary smooth k-form on Y. Let x denote any point of X. It suffices to prove that

$$[(d \circ f^{\#})(\xi)](x) = [(f^{\#} \circ d)(\xi)](x).$$

Let $y = f(x)$. As in the proof of (5.2), we can construct a smooth k-form

$$\eta = \Sigma \eta_{(i_1, \ldots, i_k)} \wedge d(\psi_{i_1}) \wedge \cdots \wedge d(\psi_{i_k})$$

on Y such that

$$\eta(v) = \xi(v)$$

holds for every point v of an open neighborhood V of y in Y. By (ED5) in (3.3), this implies that

$$[(d \circ f^{\#})(\eta)](x) = [(d \circ f^{\#})(\xi)](x)$$
$$[(f^{\#} \circ d)(\eta)](x) = [(f^{\#} \circ d)(\xi)](x).$$

Therefore, it remains to establish that

$$(d \circ f^{\#})(\eta) = (f^{\#} \circ d)(\eta).$$

Since $f^{\#}$ is an algebra homomorphism,

$$f^{\#}(\eta) = \Sigma f^{\#}[\eta_{(i_1, \ldots, i_k)}] \wedge f^{\#}[d(\psi_{i_1})] \wedge \cdots \wedge f^{\#}[d(\psi_{i_k})].$$

By the two special cases which have been established, we have

$$d\{f^{\#}[\eta_{(i_1, \ldots, i_k)}]\} = f^{\#}\{d[\eta_{(i_1, \ldots, i_k)}]\}$$
$$d\{f^{\#}[d(\psi_j)]\} = f^{\#}\{d[d(\psi_j)]\} = 0.$$

By a finite induction on the integer k, we obtain

$$\begin{aligned} d[f^{\#}(\eta)] &= \Sigma d\{f^{\#}[\eta_{(i_1,\ldots,i_k)}]\} \wedge f^{\#}[d(\psi_{i_1})] \wedge \cdots \wedge f^{\#}[d(\psi_{i_k})] \\ &= \Sigma f^{\#}\{d[\eta_{(i_1,\ldots,i_k)}]\} \wedge f^{\#}[d(\psi_{i_1})] \wedge \cdots \wedge f^{\#}[d(\psi_{i_k})] \\ &= f^{\#}\{\Sigma d[\eta_{(i_1,\ldots,i_k)}] \wedge d(\psi_{i_1}) \wedge \cdots \wedge d(\psi_{i_k})\} \\ &= f^{\#}[d(\eta)]. \end{aligned}$$

This completes the proof of (5.3).▌

Lemma 5.4. *For every integer k, the linear transformation*

$$f^{\#} : \mathbf{C}^k(Y) \to \mathbf{C}^k(X)$$

sends $\mathbf{Z}^k(Y)$ into $\mathbf{Z}^k(X)$ and $\mathbf{B}^k(Y)$ into $\mathbf{B}^k(X)$.

PROOF. Let $\xi \in \mathbf{Z}^k(Y)$ be arbitrarily given. Then we have $d(\xi) = 0$. By (5.3), we obtain

$$d[f^{\#}(\xi)] = f^{\#}[d(\xi)] = f^{\#}(0) = 0.$$

This implies $f^{\#}(\xi) \in \mathbf{Z}^k(X)$. Hence $f^{\#}$ sends $\mathbf{Z}^k(Y)$ into $\mathbf{Z}^k(X)$.

Let $\xi \in \mathbf{B}^k(Y)$ be arbitrarily given. Then there exists an $\eta \in \mathbf{C}^{k-1}(Y)$ with $d(\eta) = \xi$. By (5.3), we have

$$f^{\#}(\xi) = f^{\#}[d(\eta)] = d[f^{\#}(\eta)].$$

This implies $f^{\#}(\xi) \in \mathbf{B}^k(X)$. Hence $f^{\#}$ sends $\mathbf{B}^k(Y)$ into $\mathbf{B}^k(X)$.▌

Because of (5.4), the linear transformation

$$f^{\#} : \mathbf{C}^k(Y) \to \mathbf{C}^k(X)$$

induces a linear transformation

$$f^* = \mathbf{H}^k(f) : \mathbf{H}^k(Y) \to \mathbf{H}^k(X)$$

which will be referred to as the *induced homomorphism* of the smooth map f on the k-dimensional de Rham cohomology group $\mathbf{H}^k(Y)$.

Passing to the direct sum, we obtain a linear transformation

$$f^* = \mathbf{H}^*(f) : \mathbf{H}^*(Y) \to \mathbf{H}^*(X).$$

The following theorem is a direct consequence of (5.1).

Theorem 5.5. *The linear transformation*

$$f^* = \mathbf{H}^*(f) : \mathbf{H}^*(Y) \to \mathbf{H}^*(X)$$

is a homomorphism of the algebra $\mathbf{H}^*(Y)$ into the algebra $\mathbf{H}^*(X)$ with the following two properties:

(a) *$\mathbf{H}^*(f)$ is homogeneous of degree* 0; *that is, it sends $\mathbf{H}^k(Y)$ into $\mathbf{H}^k(X)$ for every integer k.*

(b) *$\mathbf{H}^*(f)$ preserves multiplicative unit; that is, it sends the unit of $\mathbf{H}^*(Y)$ to the unit of $\mathbf{H}^*(X)$.*

This homomorphism $f^* = \mathbf{H}^*(f)$ will be called the *induced homomorphism* of the smooth map $f : X \to Y$ on the De Rahm cohomology algebra $\mathbf{H}^*(Y)$.

The functorial properties of the induced homomorphisms given in the following two propositions are obvious from the definition.

Proposition 5.6. *The induced homomorphism*

$$i^* = \mathbf{H}^*(i) : \mathbf{H}^*(X) \to \mathbf{H}^*(X)$$

of the identity map $i : X \to X$ on a smooth n-manifold X is the identity automorphism of the algebra $\mathbf{H}^*(X)$.

Proposition 5.7. *For arbitrary smooth maps*

$$f : X \to Y \qquad g : Y \to Z$$

of smooth manifolds X, Y, and Z, we have

$$\mathbf{H}^*(g \circ f) = \mathbf{H}^*(f) \circ \mathbf{H}^*(g).$$

As a consequence of (5.6) and (5.7), we shall deduce the following corollary:

Corollary 5.8. *If $f : X \to Y$ is a diffeomorphism, then the induced homomorphism*

$$\mathbf{H}^*(f) : \mathbf{H}^*(Y) \to \mathbf{H}^*(X)$$

is an isomorphism.

PROOF. Since $f : X \to Y$ is a diffeomorphism, its inverse

$$g = f^{-1} : Y \to X$$

is defined and is a smooth map. Since the compositions $g \circ f$ and $f \circ g$ are the identity maps on X and Y, respectively, it follows that the compositions

$$\mathbf{H}^*(f) \circ \mathbf{H}^*(g) \qquad \mathbf{H}^*(g) \circ \mathbf{H}^*(f)$$

of the induced homomorphisms

$$\mathbf{H}^*(f) : \mathbf{H}^*(Y) \to \mathbf{H}^*(X) \qquad \mathbf{H}^*(g) : \mathbf{H}^*(X) \to \mathbf{H}^*(Y)$$

are the identity automorphisms of $\mathbf{H}^*(X)$ and $\mathbf{H}^*(Y)$ respectively. Hence $\mathbf{H}^*(f)$ is an isomorphism with $\mathbf{H}^*(g)$ as its inverse.∎

EXERCISES

5A. A smooth m-manifold X is said to be a *submanifold* of a smooth n-manifold Y iff X is a subspace of the topological space Y and the inclusion map $i : X \to Y$ is an immersion in the strong sense as defined at the end of (I, §5). A submanifold X of a smooth n-manifold Y is said to be a *smooth retract* of Y iff there exists a smooth map

$$r : Y \to X$$

such that $r(x) = x$ holds for every point x of X. Let X be a smooth retract of Y. Prove that the induced homomorphism

$$\mathbf{H}^*(i) : \mathbf{H}^*(Y) \to \mathbf{H}^*(X)$$

of the inclusion map $i : X \to Y$ is an epimorphism and that $\mathbf{H}^*(X)$ is isomorphic to a subalgebra of $\mathbf{H}^*(Y)$ which is a direct summand of the linear space $\mathbf{H}^*(Y)$.

5B. Consider an arbitrary smooth map

$$f : R^m \to R^n.$$

Composing f with the coordinate functions in R^n, we obtain n smooth functions

$$f_1, \ldots, f_n : R^m \to R$$

such that

$$f(p) = [f_1(p), \ldots, f_n(p)] \in R^n$$

for every point $p \in R^m$. Prove the following three statements for the induced homomorphism

$$f^{\#} : \mathbf{C}^*(R^n) \to \mathbf{C}^*(R^m).$$

(a) For every 0-form $\xi : R^n \to R$ on R^n, we have

$$f^{\#}(\xi) = \xi \circ f : R^m \to R.$$

(b) Let $x_1, \ldots, x_m$ and $y_1, \ldots, y_n$ denote the coordinate functions in R^m and R^n respectively. For every $j = 1, 2, \ldots, n$, we have

$$f^{\#}(dy_j) = \sum_{i=1}^{m} \frac{\partial f_j}{\partial x_i} dx_i.$$

(c) $f^{\#} : \mathbf{C}^*(R^n) \to \mathbf{C}^*(R^m)$ is completely determined by (a) and (b).

For example,

$$\begin{aligned} f^{\#}(dy_1 \wedge dy_2) &= f^{\#}(dy_1) \wedge f^{\#}(dy_2) \\ &= \left(\sum_{i=1}^{m} \frac{\partial f_1}{\partial x_i}\, dx_i\right) \wedge \left(\sum_{j=1}^{m} \frac{\partial f_2}{\partial x_j}\, dx_j\right) \\ &= \sum_{i<j} \left(\frac{\partial f_1}{\partial x_i}\frac{\partial f_2}{\partial x_j} - \frac{\partial f_1}{\partial x_j}\frac{\partial f_2}{\partial x_i}\right)(dx_i \wedge dx_j) \\ &= \sum_{i<j} \frac{\partial(f_1,f_2)}{\partial(x_i,x_j)}\,(dx_i \wedge dx_j). \end{aligned}$$

5C. Consider an arbitrary smooth function

$$f : R^n \to R^n$$

and let $f_1, \ldots, f_n : R^n \to R$ denote the n smooth functions such that

$$f(p) = [f_1(p), \ldots, f_n(p)] \in R^n$$

for every point $p \in R^n$. Prove that

$$f^{\#}(dx_1 \wedge \cdots \wedge dx_n) = \frac{\partial(f_1, \ldots, f_n)}{\partial(x_1, \ldots, x_n)}\, dx_1 \wedge \cdots \wedge dx_n.$$

5D. Consider the smooth map $f : R^2 \to R^2$ defined by

$$f(x,y) = (xy, 1)$$

for every point (x,y) of R^2. Compute $f^{\#}(dx)$, $f^{\#}(dy)$, and $f^{\#}(dx \wedge dy)$.

5E. Investigate possible generalizations of the concepts of the results given in this section to the case where the smooth manifolds X and Y may be infinite-dimensional.

6. Poincaré's Lemma

A subset U of the n-dimensional Euclidean space R^n is said to be *star-shaped* iff there exists a point $x_0 \in U$ such that, for every point $x \in U$, the line-segment joining x_0 and x is contained in U. The point x_0 is called a *center* of the star-shaped set U.

For example, every convex subset U of R^n is star-shaped with any point x_0 of U as center.

Theorem 6.1 (Poincaré's Lemma). *If U is a star-shaped open subset of the n-dimensional Euclidean space R^n, then*

$$\mathbf{H}^k(U) = 0$$

for every integer $k > 0$.

PROOF. According to Examples (a) and (c) of (I, §2), U is a smooth n-manifold with the inclusion map of U into R^n as a smooth chart. As in calculus, we shall use $x_1, \ldots, x_n$ to denote the coordinate functions for this chart. Because of (5.8), we may assume that the origin O of R^n is a center of the star-shaped open subset U of R^n.

For every integer $k > 0$, we shall construct a linear transformation

$$h_k : \mathbf{C}^k(U) \to \mathbf{C}^{k-1}(U).$$

For this purpose, let us first consider a smooth k-form on U of the form

$$\xi = f \wedge d(x_{i_1}) \wedge \cdots \wedge d(x_{i_k})$$

where $f : U \to R$ is a smooth function. Define a smooth function

$$g : U \to R$$

by taking

$$g(x) = \int_0^1 t^{k-1} f(tx)\, dt$$

for every point x in U. Define a smooth $(k - 1)$-form η on U by taking

$$\eta = \sum_{j=1}^{k} (-1)^{j-1} x_{i_j} \wedge d(x_{i_1}) \wedge \cdots \wedge \widehat{d(x_{i_j})} \wedge \cdots \wedge d(x_{i_k})$$

where the circumflex over $d(x_{i_j})$ indicates that $d(x_{i_j})$ is omitted. Then

$$\xi^* = g \wedge \eta$$

is a smooth $(k - 1)$-form on U.

According to (2.4), the linear space $\mathbf{C}^k(U)$ is spanned by the smooth k-forms of the form

$$\xi = f \wedge d(x_{i_1}) \wedge \cdots \wedge d(x_{i_k}).$$

Hence the assignment $\xi \to \xi^* = h_k(\xi)$ defines a linear transformation

$$h_k : \mathbf{C}^k(U) \to \mathbf{C}^{k-1}(U).$$

This completes the construction of the linear transformation h_k for every $k > 0$.

Next let us consider the following diagram of linear transformation

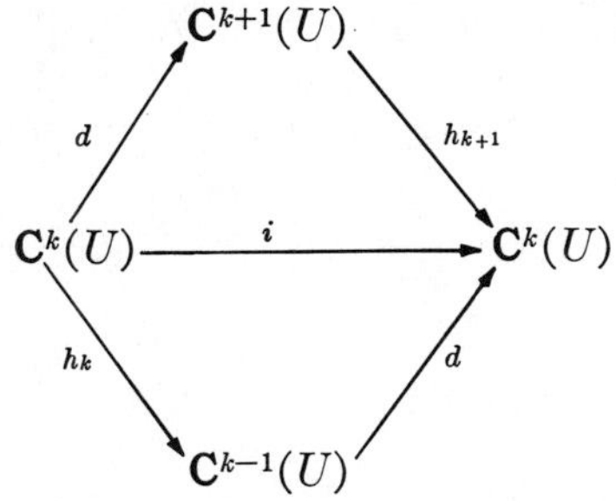

where i denotes the identity transformation on $\mathbf{C}^k(U)$. We shall establish

$$h_{k+1} \circ d + d \circ h_k = i.$$

For this purpose, we must show that

$$(h_{k+1} \circ d)(\xi) + (d \circ h_k)(\xi) = \xi$$

holds for every smooth k-form ξ on U. Let

$$\sigma = (h_{k+1} \circ d)(\xi) \qquad \tau = (d \circ h_k)(\xi).$$

We shall prove

$$\sigma + \tau = \xi.$$

Since $\mathbf{C}^k(U)$ is spanned by the smooth k-forms of the form

$$f \wedge d(x_{i_1}) \wedge \cdots \wedge d(x_{i_k})$$

we may assume

$$\xi = f \wedge d(x_{i_1}) \wedge \cdots \wedge d(x_{i_k}).$$

For every $j = 1, 2, \ldots, n$, let

$$f_j = \frac{\partial f}{\partial x_j} \qquad g_j = \frac{\partial g}{\partial x_j}.$$

Since $h_k(\xi) = \xi^* = g \wedge \eta$ and

$$d(g) = \sum_{j=1}^{n} g_j \wedge d(x_j) \qquad d(\eta) = kd(x_{i_1}) \wedge \cdots \wedge d(x_{i_k})$$

we obtain

$$\tau = d(g \wedge \eta) = \sum_{j=1}^{n} g_j \wedge d(x_j) \wedge \eta \;+\; kg \wedge d(x_{i_1}) \wedge \cdots \wedge d(x_{i_k}).$$

For every point $x \in U$, we have

$$g_j(x) = \frac{\partial}{\partial x_j} \int_0^1 t^{k-1} f(tx)\, dt = \int_0^1 t^k f_j(tx)\, dt.$$

On the other hand, since

$$d(\xi) = d(f) \wedge d(x_{i_1}) \wedge \cdots \wedge d(x_{i_k})$$
$$= \sum_{j=1}^{n} f_j \wedge d(x_j) \wedge d(x_{i_1}) \wedge \cdots \wedge d(x_{i_k})$$

we obtain

$$\sigma = h_{k+1}[d(\xi)]$$
$$= \sum_{j=1}^{n} g_j \wedge [x_j \wedge d(x_{i_1}) \wedge \cdots \wedge d(x_{i_k}) - d(x_j) \wedge \eta].$$

Consequently, we obtain

$$\sigma + \tau = \left[kg + \sum_{j=1}^{n} x_j g_j\right] \wedge d(x_{i_1}) \wedge \cdots \wedge d(x_{i_k}).$$

For every point x of U, we have

$$\left[kg + \sum_{j=1}^{n} x_j g_j\right](x) = \int_0^1 \left[kt^{k-1}f(tx) + \sum_{j=1}^{n} t^k f_j(tx)x_j\right] dt$$
$$= \int_0^1 \left[kt^{k-1}f(tx) + t^k \frac{d}{dt} f(tx)\right] dt$$
$$= \int_0^1 \frac{d}{dt}[t^k f(tx)]\, dt$$
$$= [t^k f(tx)]_0^1 = f(x).$$

This implies

$$\sigma + \tau = \xi$$

and establishes the relation

$$h_{k+1} \circ d + d \circ h_k = i$$

for every integer $k > 0$. The proof of (6.1) will be completed by the following lemma:

Lemma 6.2. *For any smooth n-manifold X and any integer $k > 0$, we have*

$$\mathbf{H}^k(X) = 0$$

if there exist two linear transformations

$$h_k : \mathbf{C}^k(X) \to \mathbf{C}^{k-1}(X) \qquad h_{k+1} : \mathbf{C}^{k+1}(X) \to \mathbf{C}^k(X)$$

such that

$$h_{k+1} \circ d + d \circ h_k$$

is the identity transformation of $\mathbf{C}^k(X)$.

PROOF. It suffices to prove that

$$\mathbf{Z}^k(X) = \mathbf{B}^k(X).$$

For this purpose, let ξ denote any closed smooth k-form on X. Then we have $d(\xi) = 0$. Hence we obtain

$$\xi = h_{k+1}[d(\xi)] + d[h_k(\xi)] = d[h_k(\xi)].$$

Therefore, ξ is exact. This proves that

$$\mathbf{Z}^k(X) \subset \mathbf{B}^k(X).$$

Because of (4.3), this implies that

$$\mathbf{Z}^k(X) = \mathbf{B}^k(X)$$

and completes the proof of (6.2).▌

EXERCISES

6A. A smooth n-manifold U is said to be *smoothly contractible* iff there exist a point $x_0 \in U$, an open interval V containing the closed unit interval $I = [0,1]$, and a smooth map

$$h : U \times V \to U$$

satisfying

$$h(x,1) = x \qquad h(x,0) = x_0$$

for every $x \in U$. Generalize the proof of (6.1) to prove

$$\mathbf{H}^k(U) = 0$$

for every $k > 0$ and every smoothly contractible open subset U of a smooth n-manifold X contained in some coordinate neighborhood of X.

6B. From Poincaré's lemma, deduce that, for any smooth function

$$f : U \to R$$

defined on any open subset U of the real line R, there exists a smooth function

$$g : U \to R$$

such that $g' = f$.

6C. From Poincaré's lemma, deduce the following assertions for any star-shaped open subset U of the Euclidean plane R^2:

(a) If $\xi, \eta : U \to R$ are two smooth functions such that

$$\frac{\partial \eta}{\partial x} - \frac{\partial \xi}{\partial y} = 0$$

then there exists a smooth function $f : U \to R$ such that

$$\frac{\partial f}{\partial x} = \xi \qquad \frac{\partial f}{\partial y} = \eta.$$

(b) For every smooth function $f : U \to R$, there exists a pair of smooth functions $\xi, \eta : U \to R$ such that

$$\frac{\partial \eta}{\partial x} - \frac{\partial \xi}{\partial y} = f.$$

6D. From Poincaré's lemma, deduce the following assertions for any star-shaped open subset U of the Euclidean space R^3:

(a) If $v : U \to R^3$ is a smooth vector field such that

$$\text{curl}(v) = 0$$

then there exists a smooth function $f : U \to R$ such that

$$\text{grad}(f) = v.$$

(b) If $v : U \to R^3$ is a smooth vector field such that

$$\text{div}(v) = 0$$

then there exists a smooth vector field $w : U \to R^3$ such that

$$\text{curl}(w) = v.$$

(c) For every smooth function $f : U \to R$, there exists a smooth vector field $v : U \to R^3$ such that

$$\text{div}(v) = f.$$

6E. Investigate possible generalizations of (6.1) to star-shaped domains in an infinite-dimensional Banach space.

CHAPTER III

Riemannian Manifolds

An introduction to Riemannian manifolds will be given in the present chapter to cover the materials needed in the proof of de Rham's theorem. For this purpose, we have to mention a few more concepts from manifold theory—for example, smooth curves and brackets of vector fields. Other results in Riemannian geometry, except a few formulated as exercises, are avoided. We shall also disregard the summation convention with repeated superscript and subscript; for example, we shall not use

$$a^i x_i$$

to denote

$$\sum_{i=1}^{n} a^i x_i.$$

1. Inner Products

Throughout the present section, let V denote an arbitrary n-dimensional linear space over R.

By an *inner product* in V, we mean a bilinear function

$$\phi : V \times V \to R$$

satisfying the following two conditions:

(IP1) ϕ is *symmetric;* that is,

$$\phi(x,y) = \phi(y,x)$$

holds for any two vectors x and y in V.

(IP2) ϕ is *positive definite;* that is,

$$\phi(x,x) > 0$$

holds for every non-zero vector x in V.

To complete (IP2), we observe that

$$\phi(0,0) = 0$$

is a direct consequence of the bilinearity of ϕ.

To establish the existence of an inner product in V, let us consider an arbitrary basis

$$B = \{e_1, \ . \ . \ . \ ,e_n\}$$

in V. Then any two vectors x and y in V can be uniquely expressed as linear combinations

$$x = \sum_{i=1}^{n} x_i e_i \qquad y = \sum_{i=1}^{n} y_i e_i$$

of B with real coefficients x_i, y_i, $(i = 1, 2, \ . \ . \ . \ , n)$. Define a function

$$\phi : V \times V \to R$$

by taking

$$\phi(x,y) = \sum_{i=1}^{n} x_i y_i$$

for any two vectors x and y in V. One can easily verify that ϕ is bilinear and satisfies the conditions (IP1) and (IP2). Hence ϕ is an inner product in V. This inner product ϕ in V will be referred to as the *canonical inner product* with respect to the basis B of V.

Let us use the symbol

$$\mathcal{I}(V)$$

to denote the set of all inner products in V. Then $\mathcal{I}(V)$ is a nonempty subset of the linear space

$$\mathcal{L}^2(V)$$

of all real-valued bilinear functions on $V \times V$. $\mathcal{I}(V)$ is not a linear subspace of $\mathcal{L}^2(V)$ since the zero function on $V \times V$ is not an inner product in V because of (IP2). However, we have the following lemma, which is obvious.

Lemma 1.1. *As a subset of $\mathcal{L}^2(V)$, $\mathcal{I}(V)$ has the following two properties:*

(a) *$\mathcal{I}(V)$ is closed under addition; that is, for any two inner products ϕ and ψ in V, the bilinear function $\phi + \psi$ is an inner product in V.*

(b) *$\mathcal{I}(V)$ is closed under multiplication by positive scalars; that is, for any inner product ϕ in V and any positive real number a, the bilinear function $a\phi$ is an inner product in V.*

Because of (a) and (b), the set $\mathcal{I}(V)$ of all inner products in V is a *cone* in the linear space $\mathcal{L}^2(V)$ of all real-valued bilinear functions on $V \times V$. The following corollary of (1.1) is an obvious consequence of the two properties (a) and (b) of $\mathcal{I}(V)$.

Corollary 1.2. *$\mathcal{I}(V)$ is a convex subset of the linear space $\mathcal{L}^2(V)$; that is, for any finite set of inner products*

$$\{\phi_1, \phi_2, \ldots, \phi_k\}$$

in V and any k non-negative real numbers $a_1, a_2, \ldots, a_k$ satisfying

$$\sum_{i=1}^{k} a_i = 1$$

the bilinear function

$$\phi = \sum_{i=1}^{k} a_i \phi_i$$

on $V \times V$ is an inner product in V.

Next let us assume that a basis

$$B = \{e_1, \ldots, e_n\}$$

of V has been arbitrarily given. Then every inner product

$$\phi : V \times V \to R$$

in V determines an $n \times n$ matrix

$$M_\phi = \begin{Vmatrix} c_{11}, & \ldots, & c_{1n} \\ \cdots & \cdots & \cdots \\ c_{n1}, & \ldots, & c_{nn} \end{Vmatrix}$$

of real numbers

$$c_{ij} = \phi(e_i,e_j) \qquad (i,j = 1, 2, \ldots, n).$$

Because of (IP1) and (IP2), the matrix M_ϕ is *symmetric* and *positive definite* in the usual sense of linear algebra. This matrix M_ϕ will be referred to as the *matrix of the inner product* ϕ with respect to the basis B of V.

Because of the bilinearity of ϕ, M_ϕ completely determines ϕ. Let

$$B^* = \{e_1^*, \ldots, e_n^*\}$$

denote the basis of the dual linear space V^* of V that is dual to B. Then we have the following proposition:

Proposition 1.3. *For any two vectors x and y in V,*

$$\phi(x,y) = \sum_{i,j=1}^{n} c_{ij}e_i^*(x)e_j^*(y)$$

where c_{ij} denotes the element of the matrix M_ϕ at the ith row and the jth column.

PROOF. Since $B = \{e_1, \ldots, e_n\}$ is a basis of V, x and y can be uniquely expressed as linear combinations

$$x = \sum_{i=1}^{n} x_ie_i \qquad y = \sum_{j=1}^{n} y_je_j$$

of B. Since $B^* = \{e_1^*, \ldots, e_n^*\}$ is dual to B, we have

$$e_i^*(x) = x_i \qquad e_j^*(y) = y_j$$

for every $i = 1, 2, \ldots, n$ and every $j = 1, 2, \ldots, n$. Then it follows from the bilinearity of ϕ that we have

$$\phi(x,y) = \sum_{i,j=1}^{n} x_iy_j\phi(e_i,e_j) = \sum_{i,j=1}^{n} c_{ij}e_i^*(x)e_j^*(y).$$

This completes the proof of (1.3).∎

Conversely, let

$$M = \begin{Vmatrix} a_{11}, & \ldots, & a_{1n} \\ \cdots & \cdots & \cdots \\ a_{n1}, & \ldots, & a_{nn} \end{Vmatrix}$$

denote any positive definite, symmetric $n \times n$ matrix of real numbers. One can easily see that the function

$$\phi : V \times V \to R$$

defined by

$$\phi(x,y) = \sum_{i,j=1}^{n} a_{ij}e_i^*(x)e_j^*(y)$$

for any two vectors x and y in V, is an inner product in V with

$$M_\phi = M.$$

The following proposition is now clear.

Proposition 1.4. *The assignment $\phi \to M_\phi$ defines an isomorphism of the semi-linear space $\mathcal{I}(V)$ onto the semi-linear space $\mathfrak{M}_n$ of all positive definite and symmetric $n \times n$ matrices of real numbers.*

Now let us assume that an inner product

$$\phi : V \times V \to R$$

in V has been arbitrarily given. In this case, the real number

$$\phi(x,y)$$

for any two given vectors x, $y \in V$ is customarily denoted by the symbol

$$\langle x,y \rangle$$

and is called the *inner product* of the vectors x and y.

By the *length* of a vector $x \in V$ (with respect to the given inner product), we mean the non-negative real number $|x|$ determined by

$$|x|^2 = \langle x,x \rangle.$$

Hence we have $|0| = 0$ and $|x| > 0$ for every nonzero vector x in V. By a *unit vector* in V, we mean a vector $x \in V$ with $|x| = 1$.

Two vectors x and y in V are said to be *orthogonal* iff

$$\langle x,y \rangle = 0.$$

From the bilinearity of the inner product, it follows that the zero vector 0 in V is orthogonal to every vector in V.

A basis $B = \{e_1, \ldots, e_n\}$ of V is said to be *orthonormal* iff $e_1, \ldots, e_n$ are mutually orthogonal unit vectors in V. The existence of an orthonormal basis of V was established in linear algebra by means of the Gram-Schmidt orthogonalization process. Now let

$$B = \{e_1, \ldots, e_n\}$$

denote any orthonormal basis of V with respect to the given inner product. Then we have the following proposition:

Proposition 1.5. *The given inner product ϕ in V is the canonical inner product with respect to the basis B.*

PROOF. Let x and y be any two vectors in V. Since B is a basis of V, x and y can be uniquely expressed as linear combinations

$$x = \sum_{i=1}^{n} x_i e_i \qquad y = \sum_{j=1}^{n} y_j e_j.$$

Since B is orthonormal, we obtain

$$\langle x,y\rangle = \sum_{i,j=1}^{n} x_i y_j \langle e_i,e_j\rangle = \sum_{k=1}^{n} x_k y_k.$$

This proves that ϕ is the canonical inner product with respect to B.▮

The assignment $x \rightarrow |x|$ defines a function

$$\nu : V \rightarrow R$$

which will be called the *norm* of V defined by the given inner product. This is justified by the following proposition, which is a direct consequence of (1.5) and the Schwarz lemma.

Proposition 1.6. *The function $\nu : V \rightarrow R$ has the following two properties:*

(a) *ν is homogeneous; that is, for every vector x in V and every real number a, we have*

$$|ax| = |a| \cdot |x|$$

where $|a|$ stands for the absolute value of the real number a.

(b) *ν satisfies the triangle inequality; that is, for any two vectors x and y in V, we have*

$$|x + y| \leq |x| + |y|.$$

Finally, the given inner product ϕ in V defines a function

$$\iota : V \rightarrow V^*$$

from the linear space V into its dual linear space V^* of all linear functionals on V by taking

$$[\iota(x)](y) = \langle x,y\rangle$$

for every $x \in V$ and every $y \in V$.

Proposition 1.7. *The function $\iota: V \to V^*$ is a bijective linear transformation of V onto V^*—in other words, an isomorphism.*

PROOF. That ι is linear follows clearly from the bilinearity of the inner product. For any nonzero vector x in V, we have

$$[\iota(x)](x) = \langle x,x \rangle \neq 0.$$

Hence we obtain $\iota(x) \neq 0$. This implies that ι is injective. Since V and V^* are both of dimension n, it follows that ι must be bijective.▌

This isomorphism $\iota : V \approx V^*$ will be called the *isomorphism defined by the given inner product* ϕ in V.

EXERCISES

1A. Review the Gram-Schmidt orthogonalization process from your favorite linear algebra book and reproduce the details of the existence proof of an orthonormal basis of V with respect to a given inner product.

1B. Give a detailed proof of (1.6).

1C. Investigate possible generalizations of the concepts and the results in this section to infinite-dimensional linear spaces.

2. Riemannian Structures

In the present section, we are concerned with an arbitrary smooth n-manifold X.

At every point x of X, let us consider the tangent space $T(X,x)$ and the semi-linear space $\mathcal{I}[T(X,x)]$ of all inner products in $T(X,x)$. Consider the set

$$\mathcal{I}(X) = \bigcup_{x \in X} \mathcal{I}[T(X,x)]$$

together with the projection

$$\pi : \mathcal{I}(X) \to X$$

defined by $\pi(\xi) = x$ for every $\xi \in \mathcal{I}[T(X,x)]$.

By a *Riemannian structure* in X, we mean a function

$$\theta : X \to \mathcal{I}(X)$$

that satisfies the following two conditions:

(RS1) The composition $\pi \circ \theta$ is the identity function on X.
(RS2) For any two smooth vector fields u and v on X, the function

$$\theta_{(u,v)} : X \to R$$

defined by

$$\theta_{(u,v)}(x) = \theta(x)[u(x),v(x)]$$

for every $x \in X$, is smooth.

Hence a Riemannian structure θ in X assigns each point $x \in X$ smoothly to an inner product $\theta(x)$ in the tangent space $T(X,x)$.

The existence of Riemannian structures in X is given by the following theorem:

Theorem 2.1. *Every smooth n-manifold X has a Riemannian structure.*

PROOF. Let $\mathfrak{C}$ denote the open cover of X that consists of all smooth coordinate neighborhoods of X. According to (I,3.11) there exists a smooth partition of unity subordinate to $\mathfrak{C}$. In other words, there exist a locally finite open refinement $\mathfrak{D}$ of $\mathfrak{C}$ and an indexed family

$$\{f_V : X \to R \mid V \in \mathfrak{D}\}$$

of smooth functions satisfying the three conditions (PU1)–(PU3) in (I,§3).

Let us consider an arbitrary open set V of X in $\mathfrak{D}$. Since $\mathfrak{D}$ is a refinement of $\mathfrak{C}$, V is contained in some smooth coordinate neighborhood U of X. Since every restriction of a smooth chart on X is also a smooth chart on X, there exists a smooth chart

$$\phi : V \to R^n.$$

For every point $x \in V$, it follows from (I,4.8) that the n tangent vectors

$$\frac{\partial}{\partial \phi_1}, \ldots, \frac{\partial}{\partial \phi_n}$$

constitute a basis B_x of the tangent space $T(X,x)$. Let $\kappa_V(x)$ denote the canonical inner product in $T(X,x)$ with respect to this basis B_x. The assignment $x \to \kappa_V(x)$ defines a function

$$\kappa_V : V \to \mathcal{G}(V).$$

For every point $x \in X$, consider the linear space $\mathfrak{L}^2[T(X,x)]$. Let

$$\mathfrak{L}^2(X) = \bigcup_{x \in X} \mathfrak{L}^2[T(X,x)].$$

Then we can define a function

$$\theta_V : X \to \mathfrak{L}^2(X)$$

$$\theta_V(x) = \begin{cases} f_V(x)\kappa_V(x) & (\text{if } x \in V) \\ 0 & (\text{if } x \in X \setminus V). \end{cases}$$

Clearly we have $\theta_V(x) \in \mathcal{I}(X)$ iff $f_V(x) \neq 0$.

Since the support $S(f_V)$ is contained in V according to (PU2) in (I, §3) and since the open cover $\mathfrak{D}$ is locally finite, we may define a function

$$\theta : X \to \mathfrak{L}^2(X)$$

by taking

$$\theta(x) = \sum_{V \in \mathfrak{D}} \theta_V(x)$$

for every $x \in X$.

Let $x \in X$ be arbitrarily given. By the definition of θ_V, we have $\theta_V(x) \in \mathcal{I}(X)$ if $\theta_V(x) \neq 0$. On the other hand, it follows from (PU3) of (I, §3) that there exists a $V \in \mathfrak{D}$ with $f_V(x) > 0$, and hence $\theta_V(x) \in \mathcal{I}(X)$. This implies $\theta(x) \in \mathcal{I}(X)$. Since θ is arbitrary, this proves that θ is a function from X to $\mathcal{I}(X)$; in symbols,

$$\theta : X \to \mathcal{I}(X).$$

To prove that this function θ is a Riemannian structure in X, it remains to verify the two conditions (RS1) and (RS2).

To verify (RS1), let x denote any point of X. By the definition of θ_V,

$$\theta_V(x) \in \mathfrak{L}^2[T(X,x)]$$

for every $V \in \mathfrak{D}$. Consequently,

$$\theta(x) \in \mathcal{I}[T(X,x)].$$

This implies $\pi[\theta(x)] = x$. Since this relation is true for every $x \in X$, $\pi \circ \theta$ is the identity function on X. This proves (RS1).

To verify (RS2), let u and v denote any two smooth vector fields on X. We have to prove that the function

$$g : X \to R$$

defined by

$$g(x) = \theta(x)[u(x),v(x)]$$

for every $x \in X$, is smooth.

For this purpose, let us first prove that the function

$$g_V : X \to R$$

defined for every $V \in \mathfrak{D}$ by

$$g_V(x) = \theta_V(x)[u(x),v(x)]$$

for every $x \in X$, is smooth. Since

$$S(g_V) \subset S(f_V) \subset V$$

it suffices to show that the restriction $g_V \mid V$ is smooth.

According to (II,2.2), there exist $2n$ smooth functions

$$a_i,b_i : V \to R \qquad (i = 1, 2, \ldots, n)$$

such that we have

$$u(x) = \sum_{i=1}^{n} a_i(x) \frac{\partial}{\partial \phi_i} \qquad v(x) = \sum_{i=1}^{n} b_i(x) \frac{\partial}{\partial \phi_i}$$

for every $x \in V$. Consequently, we obtain

$$\begin{aligned} g_V(x) &= f_V(x)\{\kappa_V(x)[u(x),v(x)]\} \\ &= f_V(x) \sum_{i=1}^{n} a_i(x)\, b_i(x) \end{aligned}$$

for every $x \in V$. This proves that $g_V \mid V$ is smooth and hence g_V is smooth for every $V \in \mathfrak{D}$.

To prove that g is smooth, let x denote an arbitrary point of X. Since $\mathfrak{D}$ is locally finite, there exists an open neighborhood W of x in X that meets only a finite number of open sets in $\mathfrak{D}$, say

$$V_1, V_2, \ldots, V_k.$$

This implies that

$$\theta(x) = \sum_{i=1}^{k} \theta_{V_i}(x)$$

for every $x \in W$ and, therefore,

$$g(x) = \sum_{i=1}^{k} g_{V_i}(x)$$

for every $x \in W$. Since g_V is smooth for every $V \in \mathfrak{D}$, this implies that g is smooth in W. Since x is any point of X and smoothness is a local property, this proves that g is smooth and thereby completes the verification of (RS2).∎

From the foregoing proof, the following corollary is obvious.

Corollary 2.2. *For an arbitrary smooth chart $\phi: V \to R^n$ on X, the assignment of $x \in V$ to the canonical inner product $\kappa_V(x)$ in $T(X,x)$ defines a Riemannian structure*

$$\kappa_V : V \to \mathcal{g}(V)$$

in V, which will be referred to as the canonical Riemannian structure in V with respect to the smooth chart ϕ.

In particular, if $V = X = R^n$ and ϕ is the identity map on R^n, then this Riemannian structure in R^n will be referred to as the *canonical Riemannian structure* in R^n.

By a *Riemannian n-manifold,* we mean a smooth n-manifold X together with a given Riemannian structure

$$\theta : X \to \mathcal{g}(X).$$

Throughout the remainder of the present section, we assume that (X,θ) is an arbitrary Riemannian n-manifold.

Theorem 2.3. *For any smooth chart*

$$\phi : V \to R^n$$

on X, there exist $n \times n$ uniquely determined smooth functions

$$g_{ij} : V \to R \qquad (i,j = 1, 2, \ldots, n)$$

such that

$$[\theta(x)](u,v) = \sum_{i,j=1}^{n} g_{ij}(x) \cdot [d_x(\phi_i)](u) \cdot [d_x(\phi_j)](v)$$

holds for every $x \in V$ and every pair (u,v) of tangent vectors in $T(X,x)$. Furthermore, for every $x \in V$, the $n \times n$ matrix

$$M_x = \begin{Vmatrix} g_{11}(x), & \ldots, & g_{1n}(x) \\ \cdots & \cdots & \cdots \\ g_{n1}(x), & \ldots, & g_{nn}(x) \end{Vmatrix}$$

of real numbers is symmetric and positive definite.

PROOF. For arbitrary $i, j = 1, 2, \ldots, n$, define a function

$$g_{ij} : V \to R$$

by taking

$$g_{ij}(x) = \theta(x)\left[\frac{\partial}{\partial\phi_i}, \frac{\partial}{\partial\phi_j}\right]$$

for every $x \in V$. By means of (II,2.1) and (I,3.10), one can easily verify that g_{ij} is smooth.

Let $x \in X$ and $u, v \in T(X,x)$ be arbitrarily given. Because of (I,4.8) there are $2n$ real numbers a_i, b_j, $(i,j = 1, 2, \ldots, n)$, such that

$$u = \sum_{i=1}^{n} a_i \frac{\partial}{\partial \phi_i} \qquad v = \sum_{j=1}^{n} b_j \frac{\partial}{\partial \phi_j}.$$

According to (I,4.18), we also have

$$[d_x(\phi_i)](u) = a_i \qquad [d_x(\phi_j)](v) = b_j$$

for every $i = 1, 2, \ldots, n$ and every $j = 1, 2, \ldots, n$. Consequently, we obtain

$$\begin{aligned}[\theta(x)](u,v) &= \sum_{i,j=1}^{n} a_i b_j [\theta(x)] \left(\frac{\partial}{\partial \phi_i}, \frac{\partial}{\partial \phi_j}\right) \\ &= \sum_{i,j=1}^{n} g_{ij}(x) \cdot [d_x(\phi_i)](u) \cdot [d_x(\phi_j)](v).\end{aligned}$$

This completes the existence proof of the first part of (2.3).

To establish the uniqueness of the functions g_{ij}, let

$$g'_{ij} : V \to R$$

denote arbitrary functions such that

$$[\theta(x)](u,v) = \sum_{i,j=1}^{n} g'_{ij}(x) \cdot [d_x(\phi_i)](u) \cdot [d_x(\phi_j)](v)$$

holds for every $x \in V$ and every pair (u,v) of tangent vectors in $T(X,x)$. If we take

$$u = \frac{\partial}{\partial \phi_i} \qquad v = \frac{\partial}{\partial \phi_j}$$

we obtain

$$g_{ij}(x) = \theta(x) \left[\frac{\partial}{\partial \phi_i}, \frac{\partial}{\partial \phi_j}\right] = g'_{ij}(x)$$

for every $x \in V$ and, therefore, we have

$$g_{ij} = g'_{ij}.$$

for every $i = 1, 2, \ldots, n$ and every $j = 1, 2, \ldots, n$. This completes the proof of the first part of (2.3).

The second part of (2.3) is obvious because M_x is the matrix of the inner product $\theta(x)$ in $T(X,x)$ with respect to the basis

$$B = \left(\frac{\partial}{\partial\phi_1}, \ . \ . \ . \ , \frac{\partial}{\partial\phi_n}\right)$$

of $T(X,x)$. This completes the proof of (2.3).||

Because of (2.3), the restriction $\theta \mid V$ of the Riemannian structure θ on the smooth coordinate neighborhood V is usually denoted by

$$\theta \mid V = \sum_{i,j=1}^{n} g_{ij}\, d\phi_i\, d\phi_j.$$

EXERCISES

2A. Let $f : X \rightarrow Y$ denote an immersion in the strong sense from a smooth m-manifold X into a Riemannian n-manifold Y with Riemannian structure $\theta : Y \rightarrow \mathcal{I}(Y)$. Prove that the function $\theta^* : X \rightarrow \mathcal{I}(X)$ defined by

$$[\theta^*(x)](u,v) = \{\theta[f(x)]\}[f_*(u),f_*(v)]$$

for every $x \in X$ and every pair (u,v) of vectors in $T(X,x)$ is a Riemannian structure in X. This Riemannian structure θ^* in X is said to be *induced* by the immersion f. In particular, if f is the immersion of a submanifold X of Y, as defined in (II, Exercise 5A), then the Riemannian structure θ^* in X is said to be *inherited* from the Riemannian manifold Y. Thus every submanifold of a Riemannian is a Riemannian manifold.

2B. Let (X,θ) and (Y,ω) denote any two Riemannian manifolds. Prove that the function

$$\theta \times \omega : X \times Y \rightarrow \mathcal{I}(X \times Y)$$

defined by

$$[(\theta \times \omega)(x,y)](u,v) = [\theta(x)](u_X,v_X) + [\omega(y)](u_Y,v_Y)$$

for every $(x,y) \in X \times Y$ and every pair (u,v) of tangent vectors in

$$T[X \times Y,(x,y)] = T(X,x) \oplus T(Y,y)$$

is a Riemannian structure in $X \times Y$. Here u_X, v_X, u_Y, v_Y stand for the components of the vectors u, v in the direct summands $T(X,x)$ and $T(Y,y)$ respectively. The Riemannian manifold $(X \times Y,\ \theta \times \omega)$ is called the *product* of the Riemannian manifolds (X,θ) and (Y,ω).

2C. Consider four real-valued functions g_{11}, g_{12}, g_{21}, and g_{22} defined on the

Euclidean plane R^2 by

$$\begin{aligned} g_{11}(x_1,x_2) &= 2x_2^2 + 1 \\ g_{12}(x_1,x_2) &= g_{21}(x_1,x_2) = x_1^2x_2 + x_1 + 2x_2 \\ g_{22}(x_1,x_2) &= x_1^4 + 3x_1^2 + 2x_1x_2 + x_2^2 + 1 \end{aligned}$$

for every point (x_1,x_2) of R^2. Define a function

$$\theta : R^2 \to \mathcal{G}(R^2)$$

by taking

$$[\theta(p)](u,v) = \sum_{i=1}^{2} \sum_{j=1}^{2} g_{ij}(p) \cdot [d_p(x_i)](u) \cdot [d_p(x_j)](v)$$

for every point $p \in R^2$ and every pair of tangent vectors u, v in $T(R^2,p)$. Verify that θ is a Riemannian structure on R^2.

2D. Consider nine real-valued functions g_{ij}, $(i,j = 1, 2, 3)$, defined on the Euclidean space R^3 by

$$\begin{aligned} g_{11}(x_1,x_2,x_3) &= x_2^2 + 1 \\ g_{22}(x_1,x_2,x_3) &= 2 \\ g_{33}(x_1,x_2,x_3) &= x_1^2 + x_2^2 + 1 \\ g_{12}(x_1,x_2,x_3) &= g_{21}(x_1,x_2,x_3) = x_2 \\ g_{23}(x_1,x_2,x_3) &= g_{32}(x_1,x_2,x_3) = x_1 \\ g_{31}(x_1,x_2,x_3) &= g_{13}(x_1,x_2,x_3) = x_2 \end{aligned}$$

for every point (x_1,x_2,x_3) of R^3. Define a function

$$\theta : R^3 \to \mathcal{G}(R^3)$$

by taking

$$[\theta(p)](u,v) = \sum_{i=1}^{3} \sum_{j=1}^{3} g_{ij}(p) \cdot [d_p(x_i)](u) \cdot [d_p(x_j)](v)$$

for every point $p \in R^3$ and every pair of tangent vectors u, v in $T(R^3,p)$. Verify that θ is a Riemannian structure on R^3.

2E. Investigate possible generalizations of the concepts and the results in this section to infinite-dimensional smooth manifolds.

3. Riemannian Metric

Throughout the present section, let X denote an arbitrary connected Riemannian n-manifold with

$$\theta : X \to \mathcal{G}(X)$$

as its given Riemannian structure. By means of θ, we shall construct a distance function

$$\rho : X \times X \to R$$

which defines the given topology of X.

By a (parametrized) *smooth curve* in X, we mean a map

$$\sigma : [a,b] \to X$$

of a closed interval $[a,b]$ of real numbers into X that can be extended to a smooth map

$$\bar{\sigma} : U \to X$$

over some open interval U containing $[a,b]$. The point $\sigma(a) \in X$ is called the *initial point* of the smooth curve σ and the point $\sigma(b) \in X$ is called the *final point* of σ. We also say that σ is a smooth curve from $\sigma(a)$ to $\sigma(b)$.

Now let us consider an arbitrary smooth curve

$$\sigma : [a,b] \to X$$

in X. Choose any smooth extension

$$\bar{\sigma} : U \to X$$

of σ over some open interval U containing $[a,b]$. For every $t \in [a,b]$, $\bar{\sigma}$ induces a linear transformation

$$\bar{\sigma}_* : T(U,t) \to T[X,\sigma(t)]$$

according to (I, §5). Since d/dt is a basis of $T(U,t)$, we obtain a tangent vector

$$\sigma_*(t) = \bar{\sigma}_*\left(\frac{d}{dt}\right) \in T[X,\sigma(t)].$$

This tangent vector $\sigma_*(t)$ is defined by

$$[\sigma_*(t)](f) = \frac{d}{dt}(f \circ \bar{\sigma})$$

for every smooth function $f : X \to R$ and hence does not depend on the choice of $\bar{\sigma}$. We shall call $\sigma_*(t)$ the *tangent vector* of the curve σ at the point $\sigma(t)$.

The Riemannian structure θ in X defines an inner product $\theta(x)$ in $T(X,x)$ for every $x \in X$. Hence we may define a function

$$s_\sigma : [a,b] \to R$$

by taking

$$s_\sigma(t) = |\sigma_*(t)| = \sqrt{\langle \sigma_*(t), \sigma_*(t) \rangle}$$

for every $t \in [a,b]$. Since s_σ is clearly continuous and nonnegative, we obtain a nonnegative real number

$$|\sigma| = \int_a^b s_\sigma(t)\, dt$$

which will be called the *length* of the smooth curve σ.

Let p and q denote any two points in X. By a *piecewise smooth curve in X from p to q,* we mean a finite sequence

$$\sigma = \{\sigma_1, \ldots, \sigma_k\}$$

of smooth curves in X such that the initial point of σ_1 is p, the final point of σ_k is q, and the initial point of σ_i is the final point of σ_{i-1} for every $i = 2, 3, \ldots, k$. The nonnegative real number

$$|\sigma| = \sum_{i=1}^{k} |\sigma_i|$$

will be called the *length* of the piecewise smooth curve σ in X.

Because of the connectedness of X, we have the following lemma:

Lemma 3.1. *For any two points p and q in X, there exists a piecewise smooth curve in X from p to q.*

PROOF. Since X is connected and locally contractible, it follows that X is pathwise connected (or arcwise connected). Hence there exists a continuous map

$$\xi : I \to X$$

from the closed unit interval $I = [0,1]$ into X with

$$\xi(0) = p \qquad \xi(1) = q.$$

For every $t \in I$, there exists a smooth chart

$$\phi_t : U_t \to R^n$$

with $\xi(t) \in U_t$ and such that $\phi_t(U_t)$ is a convex open set of R^n. Then any two points in U_t can be joined by a smooth curve in U_t.

By means of the compactness of I, one can easily establish the existence of a finite sequence

$$0 = t_0 < t_1 < \cdots < t_{i-1} < t_i < \cdots < t_k = 1$$

of real numbers such that

$$\xi(t_{i-1}) \in U_{t_i}$$

holds for every $i = 1, 2, \ldots, k$. From the choice of the charts ϕ_t, it follows that there exists a smooth curve σ_i in U_{t_i} from $\xi(t_{i-1})$ to $\xi(t_i)$ for every $i = 1, 2, \ldots, k$. Hence,

$$\sigma = \{\sigma_1, \ldots, \sigma_k\}$$

is a piecewise smooth curve in X from p to q. ∎

For any two points p and q in X, let us use the symbol

$$\Omega(p,q)$$

to denote the set of all piecewise smooth curves in X from p to q. Because of (3.1), $\Omega(p,q)$ is nonempty. The nonnegative real number

$$\rho(p,q) = \text{Inf } \{|\sigma| \mid \sigma \in \Omega(p,q)\}$$

will be referred to as the *distance* from p to q. The assignment $(p,q) \to \rho(p,q)$ defines a function

$$\rho : X \times X \to R$$

which will be called the *distance function* in the Riemannian manifold X.

Lemma 3.2. *The distance function $\rho : X \times X \to R$ is a pseudo-metric in X; that is, it satisfies the following two conditions:*

(M1) THE TRIANGLE INEQUALITY: *For any three points a, b, c in X, we have*

$$\rho(a,c) + \rho(b,c) \geq \rho(a,b).$$

(M2) *For every point p in X, we have*

$$\rho(p,p) = 0.$$

PROOF. For any point p of X, the function

$$\sigma : [0,1] \to X,$$

defined on the closed unit interval $[0,1]$ by $\sigma(t) = p$ for every $t \in [0,1]$, is a smooth curve from p to p. Since $|\sigma| = 0$, it follows that

$$\rho(p,p) = 0$$

for every point p of X. This verifies the condition (M2).

By the *reverse* of a smooth curve $\sigma : [a,b] \to X$, we mean the smooth curve $\sigma' : [a,b] \to X$ defined by

$$\sigma'(t) = \sigma(a + b - t)$$

for every $t \in [a,b]$. By the *reverse* of a piecewise smooth curve $\sigma = \{\sigma_1, \ldots, \sigma_k\}$, we mean the piecewise smooth curve

$$\sigma' = \{\sigma_k', \ldots, \sigma_1'\}.$$

One can easily verify that $|\sigma'| = |\sigma|$ holds for every piecewise smooth curve in X.

For any two points q and r of X, we clearly have

$$\Omega(r,q) = \{\sigma' \mid \sigma \in \Omega(q,r)\}.$$

Since $|\sigma'| = |\sigma|$, this implies that

(a) $$\rho(r,q) = \rho(q,r).$$

Next let p, q, r denote any three points in X. Let ϵ be an arbitrary positive real number. According to the definition of $\rho(p,r)$, there exists a piecewise smooth curve $\sigma = (\sigma_1, \ldots, \sigma_k)$ from p to r with

$$|\sigma| < \rho(p,r) + \frac{\epsilon}{2}.$$

Similarly, there exists a piecewise smooth curve $\tau = (\tau_1, \ldots, \tau_l)$ from r to q with

$$|\tau| < \rho(r,q) + \frac{\epsilon}{2}.$$

Since $\omega = (\sigma_1, \ldots, \sigma_k, \tau_1, \ldots, \tau_l)$ is a piecewise smooth curve from p to q, we have

$$\rho(p,q) \leq |\omega| = |\sigma| + |\tau| < \rho(p,r) + \rho(r,q) + \epsilon.$$

Since this is true for every positive real number ϵ, we must have

(b) $$\rho(p,q) \leq \rho(p,r) + \rho(r,q).$$

Combining (a) and (b), we obtain

$$\rho(p,q) \leq \rho(p,r) + \rho(q,r).$$

This verifies the condition (M1). Hence ρ is a pseudo-metric in X.▌

This pseudo-metric $\rho : X \times X \to R$ defines a topology τ in X as follows: For any point p of X and any positive real number r, the subset

$$N_r(p) = \{x \in X \mid \rho(p,x) < r\}$$

is called the *r-neighborhood* of the point p in X with respect to the pseudo-metric ρ. Then the topology τ is the collection of all subsets U of X such that, for every point p of U, there exists a positive real number r with $N_r(p) \subset U$. This topology τ will be referred to as the *topology in X defined by ρ.*

Lemma 3.3. *The topology in X defined by the pseudo-metric ρ coincides with the given topology in X.*

PROOF. To establish the coincidence of these two topologies in X, it suffices to prove that, for every point $p \in X$, every neighborhood of p in the given topology is a neighborhood of p in the topology defined by the pseudo-metric ρ and vice versa.

For this purpose, consider any point $p \in X$ and choose a smooth chart

$$\phi : U \to R^n$$

on X with $p \in U$ and $\phi(p) = 0$. In U, we shall consider two Riemannian structures, namely, the restriction $\theta \mid U$ of the given Riemannian structure θ in X and the canonical Riemannian structure

$$\kappa : U \to \mathcal{G}(U)$$

with respect to the smooth chart ϕ. κ makes U a part of Euclidean n-space, essentially the open subspace $\phi(U)$ of R^n.

For every $x \in U$ and every tangent vector $v \in T(X,x)$, let

$$|v| = \sqrt{[\theta(x)](v,v)} \qquad \|v\| = \sqrt{[\kappa(x)](v,v)}.$$

For every smooth curve $\sigma : [a,b] \to U$, let

$$|\sigma| = \int_a^b |\sigma_*(t)|dt \qquad \|\sigma\| = \int_a^b \|\sigma_*(t)\|dt.$$

Finally, for every piecewise smooth curve $\sigma = \{\sigma_1, \ldots, \sigma_k\}$ in U, let

$$|\sigma| = \sum_{i=1}^{k} |\sigma_i| \qquad \|\sigma\| = \sum_{i=1}^{k} \|\sigma_i\|.$$

For every point $y = (y_1, \ldots, y_n)$ in R^n, let

$$\|y\| = \sqrt{y_1^2 + \cdots + y_n^2}.$$

Since $\phi(U)$ is an open subspace of R^n containing $\phi(p) = 0$, there exists a positive real number δ such that $\phi(U)$ contains the closed δ-ball

$$B_\delta = \{y \in R^n \mid \|y\| \leq \delta\}.$$

Since ϕ is a homeomorphism of U onto $\phi(U)$, it follows that the inverse image

$$K = \phi^{-1}(B_\delta)$$

is a compact neighborhood of p with respect to the given topology in X.

Consider the unit $(n-1)$-sphere S^{n-1} in R^n defined by

$$S^{n-1} = \{y \in R^n \mid \|y\| = 1\}.$$

Because of (2.3), we may define a continuous function

$$f : K \times S^{n-1} \to R$$

by taking

$$f(x,y) = \left| \sum_{k=1}^{n} y_k \frac{\partial}{\partial \phi_k} \right| = \left[\sum_{i,j=1}^{n} g_{ij}(x) y_i y_j \right]^{1/2}$$

for every $x \in K$ and every $y = (y_1, \ldots, y_n) \in S^{n-1}$. Since $K \times S^{n-1}$ is compact and $f(x,y) > 0$ for every $(x,y) \in K \times S^{n-1}$, there exist two positive real numbers m and M such that

$$m \leq f(x,y) \leq M$$

holds for every point $(x,y) \in K \times S^{n-1}$. In other words, we have

$$m\left[\sum_{k=1}^{n} y_k^2\right]^{1/2} \leq \left[\sum_{i,j=1}^{n} g_{ij}(x)y_iy_j\right]^{1/2} \leq M\left[\sum_{k=1}^{n} y_k^2\right]^{1/2}$$

for every $(x,y) \in K \times S^{n-1}$. Because of its homogeneity in $y_1, \ldots, y_n$, this holds also for every $x \in K$ and every $y \in R^n$. This implies that

$$m\|v\| \leq |v| \leq M\|v\|$$

for every tangent vector $v \in T(X,x)$ with $x \in K$, and hence

$$m\|\sigma\| \leq |\sigma| \leq M\|\sigma\|$$

for every piecewise smooth curve in K.

Now let N be any neighborhood of p with respect to the given topology of X. There exists a positive real number $\alpha < \delta$ such that N contains every point $x \in U$ with

$$\|\phi(x)\| < \alpha.$$

We shall prove that N is a neighborhood of p with respect to the topology defined by the pseudo-metric ρ. For this purpose, it suffices to show that N contains every point $q \in X$ with

$$\rho(p,q) < m\alpha.$$

To prove this, let $q \in X \setminus N$ and let σ denote any piecewise smooth curve from p to q. Then there exists a point x on σ with

$$\|\phi(x)\| \geq \alpha$$

such that the part τ of σ from p to x is in K. This implies that

$$|\sigma| \geq |\tau| \geq m\|\tau\|.$$

In a Euclidean space, the length of a piecewise smooth curve is at least equal to the distance of its end points. Hence we have

$$\|\tau\| \geq \|\phi(x)\| \geq \alpha.$$

Consequently, we obtain

$$|\sigma| \geq m\alpha.$$

Since this is true for every piecewise smooth curve σ from p to q, we have

$$\rho(p,q) \geq m\alpha.$$

This completes the proof that N is a neighborhood of p with respect to the topology defined by ρ.

Conversely, let N be any neighborhood of p with respect to the topology defined by ρ. Then there exists a positive real number β such that N contains every point $q \in X$ with

$$\rho(p,q) < \beta.$$

To prove that N is a neighborhood of p with respect to the given topology of X, it suffices to show that N contains the open set

$$W = \left\{ x \in U \mid \|\phi(x)\| < \min\left(\frac{\beta}{M}, \delta\right) \right\}.$$

For this purpose, let $x \in U$ be arbitrarily given. Consider the smooth curve

$$\sigma : [0,1] \to X$$

defined by

$$\sigma(t) = \phi^{-1}[t\phi(x)]$$

from every $t \in [0,1]$. Then σ is a smooth curve in K from p to x with

$$\|\sigma\| = \|\phi(x)\| < \frac{\beta}{M}.$$

Consequently, we obtain

$$\rho(p,x) \leq |\sigma| \leq M\|\sigma\| < \beta.$$

Hence $x \in N$. This implies $W \subset N$ and completes the proof that N is a neighborhood of p with respect to the given topology of X.▌

Since every n-manifold is a Hausdorff space according to our definition in (I, §1), we have the following corollary:

Corollary 3.4. *The pseudo-metric $\rho : X \times X \to R$ is a metric.*

PROOF. Let p and q denote any two distinct points in X. Since X is a Hausdorff space, the subset

$$U = X \setminus \{q\}$$

is an open neighborhood of the point p. Since the pseudo-metric ρ defines the given topology of X, there exists a positive real number r with

$$N_r(p) \subset U = X \setminus \{q\}.$$

This implies that the point q is not in $N_r(p)$. Therefore, we have

$$\rho(p,q) \geq r > 0.$$

Since p and q are arbitrary distinct points in X, this proves that ρ is a metric.∎

This metric $\rho : X \times X \to R$ will be referred to as the *Riemannian metric* in X defined by the Riemannian structure θ. By (3.3), ρ defines the topology of the Riemannian manifold X.

EXERCISES

3A. Prove (3.4) directly without using (3.3).

3B. Prove that the Riemannian metric $\rho : X \times X \to R$ defined by a Riemannian structure θ on a smooth manifold X is a continuous function on the product space $X \times X$.

3C. By an *isometry* from a Riemannian manifold X onto a Riemannian manifold Y, we mean a diffeomorphism

$$f : X \to Y$$

such that its differential f_* *preserves inner products;* that is, for every point x of X and any pair of tangent vectors $u,v \in T(X,x)$, we have

$$\langle f_*(u), f_*(v) \rangle = \langle u,v \rangle.$$

Prove that every isometry $f : X \to Y$ *preserves distance;* that is, for any two points a and b in X, we have

$$\rho[f(a),f(b)] = \rho(a,b).$$

Prove that the composition of any two isometries is an isometry and that the inverse of any isometry is an isometry.

3D. Investigate possible generalizations of the concepts and the results in this section to infinite-dimensional Riemannian manifolds.

4. Riemannian Connection

Throughout the present section, let X denote an arbitrary Riemannian n-manifold with

$$\theta : X \to \mathcal{I}(X)$$

as its given Riemannian structure.

Consider the algebra $\mathbf{C}(X)$ of all smooth functions

$$f : X \to R$$

and the linear space $\mathbf{V}(X)$ of all smooth tangent vector fields

$$v : X \to T(X)$$

as defined in (II, §2). Then $\mathbf{C}(X)$ operates on the left of $\mathbf{V}(X)$ as follows: For every $f \in \mathbf{C}(X)$ and every $v \in \mathbf{V}(X)$,

$$fv : X \to T(X)$$

stands for the smooth tangent vector field defined by

$$(fv)(x) = f(x)v(x)$$

for every $x \in X$. On the other hand, $\mathbf{V}(X)$ operates on the left of $\mathbf{C}(X)$ as follows: For every $v \in \mathbf{V}(X)$ and every $g \in \mathbf{C}(X)$,

$$vg : X \to R$$

stands for the smooth function defined by

$$(vg)(x) = [v(x)](g)$$

for every $x \in X$. The following lemma is obvious.

Lemma 4.1. *For any two smooth functions $f,g : X \to R$ and every smooth tangent field $v : X \to T(X)$, we have*

$$(fv)g = f(vg).$$

Because of (4.1), the smooth function

$$fvg : X \to R$$

is well defined for arbitrarily given $f,g \in \mathbf{C}(X)$ and $v \in \mathbf{V}(X)$.

By a *connection* on the smooth n-manifold X, we mean a bilinear transformation

$$D : \mathbf{V}(X) \times \mathbf{V}(X) \to \mathbf{V}(X)$$

such that, for every $f \in \mathbf{C}(X)$ and arbitrary $u,v \in \mathbf{V}(X)$, the following two conditions are satisfied:

(C1) $D(fu,v) = fD(u,v)$

(C2) $D(u,fv) = (uf)v + fD(u,v)$.

Connections on X are historically called *affine* or *linear connections* on X. For arbitrarily given $u,v \in \mathbf{V}(X)$, the smooth tangent vector field $D(u,v)$ is called the *covariant derivative* of v in the direction of u with respect to the connection D. For any given $u \in \mathbf{V}(X)$, the linear transformation

$$D_u : \mathbf{V}(X) \to \mathbf{V}(X)$$

defined by $D_u(v) = D(u,v)$ for every $v \in \mathbf{V}(X)$, is called the *covariant differentiation* of $\mathbf{V}(X)$ in the direction of u.

To study the local structure of connections on X, let us consider an arbitrary smooth chart

$$\phi : U \to R^n$$

on the smooth n-manifold X. Let

$$D : \mathbf{V}(U) \times \mathbf{V}(U) \to \mathbf{V}(U)$$

be any connection on the smooth n-manifold U. Since

$$\frac{\partial}{\partial \phi_1}, \ . \ . \ . \ , \frac{\partial}{\partial \phi_n}$$

are smooth tangent vector fields on U and

$$d(\phi_1), \ . \ . \ . \ , d(\phi_n)$$

are smooth 1-forms on U, we may define n^3 smooth functions

$$\Gamma^i_{jk} : U \to R$$

by taking

$$\Gamma^i_{jk}(x) = [d_x(\phi_i)]D\left(\frac{\partial}{\partial \phi_j}, \frac{\partial}{\partial \phi_k}\right)$$

for arbitrary $i, j, k = 1, 2, \ . \ . \ . \ , n$ and every $x \in U$. These n^3 smooth functions Γ^i_{jk} are historically called the *Christoffel symbols* and will be referred to as the *Christoffel functions* of the given connection

$$D : \mathbf{V}(U) \times \mathbf{V}(U) \to \mathbf{V}(U).$$

The following lemma is obvious.

Lemma 4.2. *For arbitrary* $j,k = 1, 2, \ . \ . \ . \ , n$, *we have*

$$D\left(\frac{\partial}{\partial \phi_j}, \frac{\partial}{\partial \phi_k}\right) = \sum_{i=1}^{n} \Gamma^i_{jk} \frac{\partial}{\partial \phi_i}.$$

Next let u and v denote any two smooth tangent fields on U. By (II,2.2), there exist $2n$ smooth functions

$$a_j, b_k : U \to R \qquad (j,k = 1, 2, \ . \ . \ . \ , n)$$

with

$$u = \sum_{j=1}^{n} a_j \frac{\partial}{\partial \phi_j} \qquad v = \sum_{k=1}^{n} b_k \frac{\partial}{\partial \phi_k}.$$

From the bilinearity of D, together with the two conditions (C1) and (C2), it follows that

$$D(u,v) = \sum_{j,k=1}^{n} D\left(a_j \frac{\partial}{\partial \phi_j}, b_k \frac{\partial}{\partial \phi_k}\right)$$
$$= \sum_{j,k=1}^{n} \left[a_j \left(\frac{\partial}{\partial \phi_j}\right) b_k\right] \frac{\partial}{\partial \phi_k} + \sum_{j,k=1}^{n} a_j b_k D\left(\frac{\partial}{\partial \phi_j}, \frac{\partial}{\partial \phi_k}\right).$$

Therefore, by an application of (4.2), we obtain the following lemma:

Lemma 4.3. *For arbitrary smooth tangent vector fields*

$$u = \sum_{j=1}^{n} a_j \frac{\partial}{\partial \phi_j} \qquad v = \sum_{k=1}^{n} b_k \frac{\partial}{\partial \phi_k}$$

we have

$$D(u,v) = \sum_{i=1}^{n} (ub_i + c_i) \frac{\partial}{\partial \phi_i}$$

with

$$c_i = \sum_{j,k=1}^{n} a_j b_k \Gamma^i_{jk}$$

for every $i = 1, 2, \ldots, n.$

Next let us establish the following lemma:

Lemma 4.4. *Every set of* n^3 *smooth functions*

$$\Gamma^i_{jk} : U \to R \qquad (i,j,k = 1, 2, \ldots, n)$$

is the set of the Christoffel functions of some connection D *on* U.

PROOF. By means of the n^3 given smooth functions Γ^i_{jk}, define a function

$$D : \mathbf{V}(U) \times \mathbf{V}(U) \to \mathbf{V}(U)$$

as follows. Let $u, v \in \mathbf{V}(U)$ be arbitrarily given. There are $2n$ uniquely determined smooth functions

$$a_j, b_k : U \to R \qquad (j,k = 1, 2, \ldots, n)$$

such that

$$u = \sum_{j=1}^{n} a_j \frac{\partial}{\partial \phi_j} \qquad v = \sum_{k=1}^{n} b_k \frac{\partial}{\partial \phi_k}.$$

Then we define D by taking

$$D(u,v) = \sum_{i=1}^{n} (ub_i + c_i) \frac{\partial}{\partial \phi_i}$$

where c_i stands for the smooth function $c_i : U \to R$ defined by

$$c_i = \sum_{j,k=1}^{n} a_j b_k \Gamma^i_{jk}$$

for every $i = 1, 2, \ldots, n$. It remains to verify that D is a connection on U.

Since D is clearly bilinear, it remains to verify the conditions (C1) and (C2). For this purpose, let $f \in \mathbf{C}(X)$ and $u, v \in \mathbf{V}(X)$ be arbitrarily given. Then we have

$$\begin{aligned}
D(fu,v) &= \sum_{i=1}^{n} \left[(fu)b_i + \sum_{j,k=1}^{n} (fa_j)b_k\Gamma^i_{jk} \right] \frac{\partial}{\partial \phi_i} \\
&= f \sum_{i=1}^{n} \left[ub_i + \sum_{j,k=1}^{n} a_j b_k \Gamma^i_{jk} \right] \frac{\partial}{\partial \phi_i} \\
&= fD(u,v) \\
D(u,fv) &= \sum_{i=1}^{n} \left[u(fb_i) + \sum_{j,k=1}^{n} a_j(fb_k)\Gamma^i_{jk} \right] \frac{\partial}{\partial \phi_i} \\
&= \sum_{i=1}^{n} (uf)\, b_i \frac{\partial}{\partial \phi_i} + f \sum_{i=1}^{n} (ub_i + c_i) \frac{\partial}{\partial \phi_i} \\
&= (uf)v + fD(u,v).
\end{aligned}$$

Hence D is a connection on U. This completes the proof of (4.4).∎

To define the meaning of a Riemannian connection on X, we shall use the concepts of the *bracket* and the *inner product* of any two smooth tangent vector fields on X.

By the *bracket* of any two smooth tangent vector fields u and v on X, we mean the smooth tangent vector field

$$[u,v] : X \to T(X)$$

determined by

$$[u,v](f) = u(vf) - v(uf)$$

for every smooth function $f : X \to R$. The assignment $(u,v) \to [u,v]$ defines a function

$$\beta : \mathbf{V}(X) \times \mathbf{V}(X) \to \mathbf{V}(X).$$

By straightforward verification, we have the following lemma:

Lemma 4.5. *The function β has the following three properties:*

(a) *β is bilinear.*

(b) *β is antisymmetric; that is, for any two smooth tangent vector fields u and v on X, we have*

$$[u,v] = -[v,u].$$

(c) *β satisfies the Jacobi identity; that is, for any three smooth tangent vector fields u, v, and w on X, we have*

$$[[u,v],w] + [[v,w],u] + [[w,u],v] = 0.$$

By the *inner product* of any two smooth vector fields u and v on X, we mean the smooth function

$$\langle u,v\rangle : X \to R$$

defined by

$$\langle u,v\rangle(x) = \langle u(x),v(x)\rangle = [\theta(x)](u,v)$$

for every $x \in X$. The assignment $(u,v) \to \langle u,v\rangle$ clearly defines a bilinear function

$$\gamma : \mathbf{V}(X) \times \mathbf{V}(X) \to \mathbf{C}(X).$$

By a *Riemannian connection* on a Riemannian n-manifold X, we mean a connection

$$D : \mathbf{V}(X) \times \mathbf{V}(X) \to \mathbf{V}(X)$$

on X such that, for arbitrary smooth tangent vector fields u, v, and w on X, the following two conditions are satisfied:

(RC1) $D(u,v) - D(v,u) = [u,v]$.
(RC2) $\langle D(w,u),v\rangle + \langle u,D(w,v)\rangle = w\langle u,v\rangle$.

Theorem 4.6. *There exists a unique Riemannian connection on any given Riemannian n-manifold X.*

PROOF. *Uniqueness.* Let D denote an arbitrary Riemannian connection on X and let

$$\phi : U \to R^n$$

be any smooth chart on X. Consider the Christoffel functions

$$\Gamma^i_{jk} : U \to R \qquad (i,j,k = 1, 2, \ldots, n)$$

of D on U together with the smooth functions

$$g_{jk} = \left\langle \frac{\partial}{\partial \phi_j}, \frac{\partial}{\partial \phi_k} \right\rangle \qquad (j,k = 1, 2, \ldots, n).$$

By an elementary property of the partial derivatives of the second order of smooth functions defined on $\phi(U)$, we have

$$\left[\frac{\partial}{\partial \phi_r}, \frac{\partial}{\partial \phi_s}\right] = 0 \qquad (r,s = 1, 2, \ldots, n).$$

In view of (RC1), these imply

$$D\left(\frac{\partial}{\partial \phi_r}, \frac{\partial}{\partial \phi_s}\right) = D\left(\frac{\partial}{\partial \phi_s}, \frac{\partial}{\partial \phi_r}\right)$$

for every $r, s = 1, 2, \ldots, n$. Then it follows from (RC2) that we have

$$\frac{\partial}{\partial \phi_j}\left\langle \frac{\partial}{\partial \phi_k}, \frac{\partial}{\partial \phi_l} \right\rangle + \frac{\partial}{\partial \phi_k}\left\langle \frac{\partial}{\partial \phi_j}, \frac{\partial}{\partial \phi_l} \right\rangle - \frac{\partial}{\partial \phi_l}\left\langle \frac{\partial}{\partial \phi_j}, \frac{\partial}{\partial \phi_k} \right\rangle = 2\left\langle D\left(\frac{\partial}{\partial \phi_j}, \frac{\partial}{\partial \phi_k}\right), \frac{\partial}{\partial \phi_l} \right\rangle.$$

Because of (4.2), this implies that

$$2\sum_{i=1}^{n} \Gamma^i_{jk} g_{il} = \frac{\partial}{\partial \phi_j} g_{kl} + \frac{\partial}{\partial \phi_k} g_{jl} - \frac{\partial}{\partial \phi_l} g_{jk}$$

for every $j, k, l = 1, 2, \ldots, n$.

Let $j, k = 1, 2, \ldots, n$ be arbitrarily fixed integers. Since the matrix $g = \|g_{il}\|$ is nonsingular for every $x \in U$, we can solve these n equations and obtain

$$\Gamma^i_{jk} = \frac{1}{2}\sum_{l=1}^{n} G_{il}\left(\frac{\partial}{\partial \phi_j} g_{kl} + \frac{\partial}{\partial \phi_k} g_{jl} - \frac{\partial}{\partial \phi_l} g_{jk}\right)$$

for every $i = 1, 2, \ldots, n$. Here G_{il} stands for the smooth function at the ith row and the lth column of the inverse matrix

$$G = g^{-1}$$

of $g = \|g_{il}\|$. This proves that Γ^i_{jk} is completely determined by the smooth chart ϕ and the given Riemannian structure θ on X. In view of (4.3), this completes the uniqueness proof.

Existence. According to (4.4), we may define for every smooth chart $\phi : U \to R^n$ on X a connection

$$D_\phi : \mathbf{V}(U) \times \mathbf{V}(U) \to \mathbf{V}(U)$$

on U by taking

$$\Gamma^i_{jk} = \frac{1}{2}\sum_{l=1}^{n} G_{il}\left(\frac{\partial}{\partial\phi_j}g_{kl} + \frac{\partial}{\partial\phi_k}g_{jl} - \frac{\partial}{\partial\phi_l}g_{jk}\right)$$

as its n^3 Christoffel functions. One can easily verify that D_ϕ satisfies the two conditions (RC1) and (RC2). Hence D_ϕ is a Riemannian connection on U.

Define a function

$$D : \mathbf{V}(X) \times \mathbf{V}(X) \to \mathbf{V}(X)$$

as follows: Let $u, v \in \mathbf{V}(X)$ be arbitrarily given. We have to define a smooth tangent vector field

$$w = D(u,v) : X \to T(X).$$

For this purpose, let x denote any point of X. Choose a smooth chart $\phi : U \to R^n$ on X with $x \in U$. Then $u \mid U$ and $v \mid U$ are smooth tangent vector fields on U. We define part of this theorem, the tangent vector

$$[D_\phi(u \mid U, v \mid V)](x) \in T(X,x)$$

does not depend on the choice of the smooth chart ϕ. Hence we may define

$$w(x) = [D_\phi(u \mid U, v \mid V)](x).$$

This completes the construction of the tangent vector field $w = D(u,v)$, which is clearly smooth. This assignment $(u,v) \to D(u,v)$ defines a function

$$D : \mathbf{V}(X) \times \mathbf{V}(X) \to \mathbf{V}(X)$$

which is clearly bilinear and satisfies the four conditions (C1), (C2), (RC1), and (RC2). Hence D is a Riemannian connection on X. This completes the proof of (4.6).▌

The following corollary is a direct consequence of (2.1) and (4.6).

Corollary 4.7. *Every smooth n-manifold has a connection.*

EXERCISES

4A. By the *torsion* of a connection D on X, we mean the bilinear function

$$T : \mathbf{V}(X) \times \mathbf{V}(X) \to \mathbf{V}(X)$$

defined by

$$T(u,v) = D(u,v) - D(v,u) - [u,v]$$

for any two smooth tangent vector fields on X. Prove that

(a) $T(u,v) = -T(v,u)$

(b) $T(fu,v) = fT(u,v) = T(u,fv)$

for arbitrary u, $v \in \mathbf{V}(X)$ and $f \in \mathbf{C}(X)$. The connection D is said to be *symmetric*, or *torsion-free*, iff $T(u,v) = 0$ holds for all u, $v \in \mathbf{V}(X)$. Show that every smooth n-manifold X has a symmetric connection.

4B. Let $\mathbf{L}[T(X,x)]$ denote the algebra of all linear transformations of the tangent space $T(X,x)$ into itself. Let

$$\mathbf{L}(X) = \bigcup_{x \in X} \mathbf{L}[T(X,x)].$$

Show that $\mathbf{L}(X)$ is an algebra. By the *curvature* of a connection D on X, we mean the bilinear function

$$R : \mathbf{V}(X) \times \mathbf{V}(X) \to \mathbf{L}(X)$$

determined by the formula

$$[R(u,v)]w = D(u,D(v,w)) - D(v,D(u,w)) - D([u,v],w)$$

for all u, v, $w \in \mathbf{V}(X)$. Prove that

(a) $R(u,v) = -R(v,u)$

(b) $R(fu,v) = fR(u,v) = R(u,fv)$

(c) $[R(u,v)](fw) = f[R(u,v)](w)$

for arbitrary u, v, $w \in \mathbf{V}(X)$ and $f \in \mathbf{C}(X)$.

4C. For an arbitrary 1-form ξ on X, prove that the exterior differential $d\xi$ of ξ has the property

$$d\xi(u,v) = \tfrac{1}{2}\{u(\xi v) - v(\xi u) - \xi[u,v]\}$$

for any smooth tangent vector fields u and v on X.

4D. Investigate possible generalizations of the concepts and the results in this section to the case in which the Riemannian manifold X is infinite-dimensional.

5. Geodesics

Throughout the present section, we are concerned with an arbitrary smooth n-manifold X together with a given connection

$$D : \mathbf{V}(X) \times \mathbf{V}(X) \to \mathbf{V}(X)$$

on X.

Consider an arbitrary smooth curve

$$\sigma : [a,b] \to X$$

in X. Let $t \in [a,b]$ be arbitrarily given. The tangent vector

$$\sigma_*(t) = \bar{\sigma}_*(d/dt) \in T[X,\sigma(t)]$$

of σ at the point $\sigma(t)$ is defined in §3. Select a smooth chart

$$\phi : U \to R^n$$

on X with $\sigma(t) \in U$. Let $V = \sigma^{-1}(U)$ and consider the smooth functions

$$f_i : V \to R \qquad (i = 1, 2, \ldots, n)$$

defined by

$$f_i(s) = \phi_i[\sigma(s)]$$

for every $s \in V$. Here $\phi_i = p_i \circ \phi$ denotes the ith coordinate function of the chart ϕ. Then we have

$$\sigma_*(t) = \sum_{i=1}^{n} f_i'(t) \frac{\partial}{\partial \phi_i}$$

where f_i' denotes the derivative of the smooth function f_i for every $i = 1, 2, \ldots, n$.

On the basis of (4.3), we may define a tangent vector

$$D[\sigma_*(t),\sigma_*(t)] = \sum_{i=1}^{n} \left\{ f_i''(t) + \sum_{j,k=1}^{n} f_j'(t) f_k'(t) \Gamma_{jk}^i[\sigma(t)] \right\} \frac{\partial}{\partial \phi_i}$$

in $T[X,\sigma(t)]$, where

$$\Gamma_{jk}^i : U \to R \qquad (i,j,k = 1, 2, \ldots, n)$$

stand for the n^3 Christoffel functions of the connection D with respect to the smooth chart ϕ.

By a *geodesic* in X with respect to the connection D, we mean a smooth curve

$$\sigma : [a,b] \to X$$

such that

$$D[\sigma_*(t),\sigma_*(t)] = 0$$

holds for every $t \in [a,b]$.

Theorem 5.1. *For every point $x \in X$ and every tangent vector $v \in T(X,x)$, there exists a positive real number r such that we have a unique geodesic*

$$\sigma : [-r,r] \to X$$

in X satisfying $\sigma(0) = x$ and $\sigma_*(0) = v$.

PROOF. Choose a smooth chart

$$\phi : U \to R^n$$

on X with $x \in U$ and let

$$\Gamma^i_{jk} : U \to R \qquad (i,j,k = 1, 2, \ldots, n)$$

denote the Christoffel functions of D with respect to this smooth chart ϕ.

Define n^3 smooth function

$$F^i_{jk} : \phi(U) \to R \qquad (i,j,k = 1, 2, \ldots, n)$$

by taking

$$F^i_{jk}(y_1, \ldots, y_n) = \Gamma^i_{jk}[\phi^{-1}(y_1, \ldots, y_n)]$$

for every point $(y_1, \ldots, y_n) \in \phi(U) \subset R^n$. Consider the system

$$\frac{d^2y_i}{dt^2} + \sum_{j,k=1}^{n} F^i_{jk}(y_1, \ldots, y_n) \frac{dy_j}{dt}\frac{dy_k}{dt} = 0 \qquad (i = 1, 2, \ldots, n)$$

of n ordinary differential equations of the second order, with the initial conditions

$$y_i(0) = \phi_i(x) \qquad (i = 1, 2, \ldots, n)$$

and

$$\sum_{i=1}^{n} y_i'(0) \frac{\partial}{\partial \phi_i} = v.$$

The theory of ordinary differential equations provides us with a positive real number r and n uniquely determined smooth functions

$$f_i : [-r,r] \to R \qquad (i = 1, 2, \ldots, n)$$

such that, for every $t \in [-r,r]$, the point

$$y = [f_1(t), \ldots, f_n(t)]$$

of R^n is in $\phi(U)$ and that the system

$$y_i = f_i(t) \qquad (i = 1, 2, \ldots, n)$$

satisfies the n differential equations as well as the initial conditions. See [Co, Chapter 6].

Define a smooth curve

$$\sigma : [-r,r] \to X$$

by taking

$$\sigma(t) = \phi^{-1}[f_1(t), \ . \ . \ . \ , f_n(t)]$$

for every $t \in [-r,r]$. Because of the initial conditions, we clearly have

$$\sigma(0) = x \qquad \sigma_*(0) = v.$$

On the other hand, since

$$f_i''(t) + \sum_{j,k=1}^{n} f_j'(t)f_k'(t)\Gamma_{jk}^i[\sigma(t)]$$
$$= f_i''(t) + \sum_{j,k=1}^{n} F_{jk}^i[f_1(t), \ . \ . \ . \ , f_n(t)]f_j'(t)f_k'(t) = 0$$

holds for every $i = 1, 2, \ . \ . \ . \ , n$ and every $t \in [-r,r]$, we have

$$D[\sigma_*(t),\sigma_*(t)] = 0$$

for every $t \in [-r,r]$. Hence σ is a geodesic. This completes the existence proof.

To prove the uniqueness, let

$$\tau : [-r,r] \to X$$

denote an arbitrary geodesic in X with

$$\tau(0) = x \qquad \tau_*(0) = v.$$

It remains to prove that $\tau(t) = \sigma(t)$ for every $t \in [-r,r]$.

Since $\tau^{-1}(U)$ is an open set of $[-r,r]$ containing 0, there exists a largest positive real number $q \leq r$ such that $\tau(t) \in U$ holds for every real number t with $|t| < q$. Let

$$g_i : (-q,q) \to R \qquad (i = 1, 2, \ . \ . \ . \ , n)$$

denote the smooth functions defined on the open interval $(-q,q)$ by

$$g_i(t) = \phi_i[\tau(t)]$$

for every $t \in (-q,q)$. Then, we have

$$D[\tau_*(t),\tau_*(t)] = \sum_{i=1}^{n} \left\{ g_i''(t) + \sum_{j,k=1}^{n} g_j'(t)g_k'(t)\Gamma_{jk}^i[\tau(t)] \right\} \frac{\partial}{\partial \phi_i}$$

for every $t \in (-q,q)$. Since τ is a geodesic, this implies that

$$g_i''(t) + \sum_{j,k=1}^{n} F_{jk}^i[g_1(t), \ldots, g_n(t)]g_j'(t)g_k'(t) = 0$$

for every $i = 1, 2, \ldots, n$ and every real number t with $|t| < q$. Hence the system

$$y_i = g_i(t) \qquad (i = 1, 2, \ldots, n)$$

satisfies the n differential equations

$$\frac{d^2y_i}{dt^2} + \sum_{j,k=1}^{n} F_{jk}^i(y_1, \ldots, y_n)\frac{dy_j}{dt}\frac{dy_k}{dt} = 0 \qquad (i = 1, 2, \ldots, n)$$

for every $t \in (-q,q)$.

On the other hand, since $\tau(0) = x$ and $\tau_*(0) = v$, the initial conditions

$$y_i(0) = \phi_i(x) \qquad (i = 1, 2, \ldots, n)$$

and

$$\sum_{i=1}^{n} y_i'(0)\frac{\partial}{\partial\phi_i} = v$$

are also satisfied by $y_i = g_i(t)$, $(i = 1, 2, \ldots, n)$. It follows from the uniqueness in the theory of ordinary differential equations that

$$g_i(t) = f_i(t)$$

holds for every integer $i = 1, 2, \ldots, n$ and every real number t with $|t| < q$. This implies that

$$\tau(t) = \sigma(t)$$

for every real number t with $|t| < q$. Since σ and τ are continuous, this implies

$$\tau(-q) = \sigma(-q) \in U \qquad \tau(q) = \sigma(q) \in U.$$

Assume $q < r$. Since U is open and τ is continuous, these imply the existence of a real number q' with $q < q' \leq r$ such that $\tau(t) \in U$ holds for every $t \in [-q',q']$. This contradicts the choice of q and proves that $q = r$. Hence we obtain

$$\tau(t) = \sigma(t)$$

for every $t \in [-r,r]$. This completes the proof of (5.1).∎

The existence and uniqueness theory of ordinary differential equations can give us much more information than the conclusion of the preceding theorem. In particular, we have the following proposition.

Consider the tangent bundle $T(X)$ of X as defined in (I,§6) and the natural projection

$$\pi: T(X) \to X.$$

For every $x \in X$, let

$$O_x \in T(X,x) \subset T(X)$$

denote the zero tangent vector of X at the point x.

By means of the existence and uniqueness theory of ordinary differential equations as in the proof of (5.1), one can easily establish the following proposition:

Proposition 5.2. *For every point $x \in X$, there exist a positive real number r_x and an open neighborhood V_x of O_x in $T(X)$ together with a smooth map*

$$M_x : [-r_x,r_x] \times V_x \to X$$

such that, for every $v \in V_x$, the smooth curve

$$\sigma_v : [-r_x,r_x] \to X$$

defined by

$$\sigma_v(t) = M_x(t,v)$$

for every $t \in [-r_x,r_x]$, is the unique geodesic in X on $[-r_x,r_x]$ and satisfying

$$\sigma_v(0) = \pi(v) \qquad \sigma_*(0) = v.$$

The following proposition follows directly from the nature of the differential equations and the initial conditions in the proof of (5.1).

Proposition 5.3. *For arbitrary $v \in V_x$ and $s,t \in R$ such that*

$$sv \in V_x \qquad |t| \leq r_x \qquad |st| \leq r_x$$

we have

$$\sigma_{sv}(t) = \sigma_v(st).$$

For every $x \in X$, the subset

$$W_x = \{v \in T(X) \mid t^{-1}v \in V_x \text{ for every } t \geq r_x\}$$

of $T(X)$ is clearly an open neighborhood of O_x in $T(X)$. Thus we obtain an open subspace

$$W = \bigcup_{x \in X} W_x$$

of $T(X)$. The following proposition is obvious.

Proposition 5.4. *For every $x \in X$, the intersection*

$$S_x = W \cap T(X,x)$$

is a star-shaped open set of $T(X,x)$ with O_x as center.

Now let us define a smooth map

$$\exp : W \to X$$

as follows: Let $v \in W$ be arbitrarily given. Then there exists a point $x \in X$ with $v \in W_x$, and hence

$$r_x^{-1} v \in V_x.$$

Because of (5.3) and the uniqueness of the geodesic, the point

$$M_x(r_x, r_x^{-1} v) = \sigma_{r_x^{-1} v}(r_x) \in X$$

does not depend on the choice of the point $x \in X$. Hence we may define exp by taking

$$\exp(v) = M_x(r_x, r_x^{-1} v).$$

This completes the definition of exp. Since M_x is smooth, it follows that the restriction $\exp \mid W_x$ is also smooth. Since smoothness is a local property, this implies that

$$\exp : W \to X$$

is a smooth map, which will be referred to as the *exponential map*. For every $x \in X$, its restriction of S_x will be denoted by

$$\exp_x : S_x \to X$$

and will be called the *exponential map* at the point $x \in X$.

For every $x \in X$, since $\exp_x$ is a smooth map sending O_x to x, it defines a differential

$$(\exp_x)_* : T(S_x, O_x) \to T(X, x)$$

as described in (I, §5).

Lemma 5.5. *For every $x \in X$, the differential*

$$(\exp_x)_* : T(S_x, O_x) \to T(X, x)$$

is bijective.

PROOF. Define a linear map

$$\lambda : T(X, x) \to T(S_x, O_x)$$

as follows: Let $v \in T(X,x)$ be arbitrarily given. Since S_x is a star-shaped open set of $T(X,x)$ with O_x as the center, there exists a positive real number r such that $tv \in S_x$ holds for every real number t with $|t| \leq r$. Then we define

$$\lambda(v) = \sigma_*(0)$$

with $\sigma_*(0)$ denoting the tangent vector of the smooth curve

$$\sigma : [-r,r] \to S_x$$

defined by $\sigma(t) = tv$ for every $t \in [-r,r]$.

Since the smooth curve

$$\tau : [-r,r] \to X$$

defined by

$$\tau(t) = \exp_x [\sigma(t)] = \exp_x (tv)$$

for every $t \in [-r,r]$, is clearly a geodesic in X with $\tau(0) = x$ and $\tau_*(0) = v$, we obtain

$$(\exp_x)_*[\lambda(v)] = v.$$

This implies that the composition $(\exp_x)_* \circ \lambda$ of the two linear transformations λ and $(\exp_x)_*$ is the identity map on $T(X,x)$. It follows that $(\exp_x)_*$ is surjective. Since the linear spaces $T(S_x,O_x)$ and $T(X,x)$ are of the same finite dimension n, this implies that $(\exp_x)_*$ is bijective.∎

The following corollary is a direct consequence of (5.5) and (I,5.9).

Corollary 5.6. *The exponential map* $\exp_x$ *at any point* $x \in X$ *sends some open neighborhood* G_x *of* O_x *in* $T(X,x)$ *diffeomorphically onto an open neighborhood* N_x *of* x *in* X.

For every $x \in X$, consider the open neighborhood N_x of x in X. We may assume that N_x is contained in the coordinate neighborhood U of some smooth chart

$$\phi : U \to R^n$$

of X. Define n smooth functions

$$\nu_i : N_x \to R \qquad (i = 1, 2, \ldots, n)$$

by taking

$$\nu_i(y) = d_y(\phi_i)[\exp_x^{-1}(y)]$$

for every $i = 1, 2, \ldots, n$ and every $y \in N_x$. Then we obtain a smooth chart

$$\nu = N_x \to R^n$$

on X by taking $\nu_1, \ldots, \nu_n$ as its coordinate functions. This smooth chart ν will be referred to as a *geodesic chart*, or a *normal chart*, on X with respect to the given connection D on X. Then open neighborhood N_x will be called a *geodesic neighborhood* of x or a *normal coordinate neighborhood* of x.

The following proposition is obvious.

Proposition 5.7. *A smooth curve*

$$\sigma : [a,b] \to N_x$$

is a geodesic in X passing through the point x iff the composition

$$\nu \circ \sigma : [a,b] \to R^n$$

is a line segment in R^n passing through the origin O of R^n.

Now let us consider the smooth map

$$h : W \to X \times X$$

defined by

$$h(v) = [\pi(v), \exp(v)]$$

for every $v \in W$. Since $h(O_x) = (x,x)$, h defines a differential

$$h_* : T(W,O_x) \to T[X \times X, (x,x)].$$

Lemma 5.8. *For every $x \in X$, the differential*

$$h_* : T(W,O_x) \to T[X \times X, (x,x)]$$

is bijective.

PROOF. The tangent space $T(W,O_x)$ can be identified with the direct sum

$$T(X,x) \oplus T(S_x,O_x)$$

by means of a natural isomorphism, while the tangent space $T[X \times X, (x,x)]$ is essentially the direct sum

$$T(X,x) \oplus T(X,x).$$

In terms of these direct sums, h_* is the identity on the first direct summand and $(\exp_x)_*$ on the second direct summand. Because of (5.5), this implies that h_* is bijective.▌

The following corollary is a direct consequence of (5.8) and (I,5.9).

Corollary 5.9. *For every point $x \in X$, the smooth map h sends some open neighborhood H_x of O_x in $T(X)$ diffeomorphically onto an open neighborhood L_x of the point (x,x) in $X \times X$.*

EXERCISES

5A. Consider an arbitrary isometry $f : X \to Y$ of a Riemannian manifold X onto a Riemannian manifold Y. Prove that, for any geodesic

$$\sigma : [a,b] \to X$$

in X, the composition

$$\tau = f \circ \sigma : [a,b] \to Y$$

is a geodesic in Y.

5B. Consider any two isometries

$$f, g : X \to Y$$

of a connected Riemannian manifold X onto a connected Riemannian manifold Y. Prove that $f = g$ holds iff there exists a point x in X such that the following two conditions are satisfied:

(a) $f(x) = g(x)$.

(b) $f_* \mid T(X,x) = g_* \mid T(X,x)$.

5C. Prove that, in a Riemannian manifold X that is complete with respect to its Riemannian metric

$$\rho : X \times X \to R$$

every pair of points p and q in X can be joined by a geodesic of length $\rho(p,q)$.

5D. Investigate possible generalizations of the concepts and the results in this section and apply them to the case in which the smooth manifold X is infinite-dimensional.

6. Convex Neighborhoods

Throughout the present section, we are concerned with an arbitrary Riemannian n-manifold X. According to (4.6), there exists a uniquely determined Riemannian connection

$$D : \mathbf{V}(X) \times \mathbf{V}(X) \to \mathbf{V}(X).$$

A subset C of X is said to be *convex* iff, for any two points, there exists a unique geodesic σ that joins the two points and lies in C. Our main objective is to establish the existence of arbitrarily small convex open neighborhoods at every point of X.

Let x denote an arbitrary point of X and consider any geodesic chart

$$\nu : N_x \to R^n$$

on X with respect to the Riemannian connection D as constructed in the preceding section. Let

$$\Gamma^i_{jk} : N_x \to R \qquad (i, j, k = 1, 2, \ldots, n)$$

denote the n^3 Christoffel functions of the connection D with respect to the smooth chart ν.

Lemma 6.1. $\Gamma^i_{jk}(x) = 0$ *holds for every* $i, j, k = 1, 2, \ldots, n$.

PROOF. Because of (4.2), it suffices to establish

$$\left[D\left(\frac{\partial}{\partial \nu_j}, \frac{\partial}{\partial \nu_k}\right)\right](x) = 0$$

for every $j, k = 1, 2, \ldots, n$.

For this purpose, let us assume that $j = k$. Since the smooth curve σ in X defined by

$$\nu_i[\sigma(t)] = \begin{cases} t & (\text{if } i = j) \\ 0 & (\text{if } i \neq j) \end{cases}$$

for every real number t with sufficiently small absolute value $|t|$ is a geodesic with tangent vector $\partial/\partial\nu_j$, we have

$$\left[D\left(\frac{\partial}{\partial \nu_j}, \frac{\partial}{\partial \nu_j}\right)\right](x) = 0.$$

Next assume $j \neq k$. Since the smooth curve σ in X defined by

$$\nu_i[\sigma(t)] = \begin{cases} t & (\text{if } i = j \text{ or } i = k) \\ 0 & (\text{if } i \neq j \text{ and } i \neq k) \end{cases}$$

for every real number t with sufficiently small absolute value $|t|$ is a geodesic with tangent vector

$$\frac{\partial}{\partial \nu_j} + \frac{\partial}{\partial \nu_k}$$

we have

$$\left[D\left(\frac{\partial}{\partial \nu_j} + \frac{\partial}{\partial \nu_k}, \frac{\partial}{\partial \nu_j} + \frac{\partial}{\partial \nu_k}\right)\right](x) = 0.$$

Since D is bilinear and since

$$\left[D\left(\frac{\partial}{\partial \nu_j}, \frac{\partial}{\partial \nu_j}\right)\right](x) = 0 \qquad \left[D\left(\frac{\partial}{\partial \nu_k}, \frac{\partial}{\partial \nu_k}\right)\right](x) = 0$$

we obtain

$$\left[D\left(\frac{\partial}{\partial\nu_j}+\frac{\partial}{\partial\nu_k},\frac{\partial}{\partial\nu_j}+\frac{\partial}{\partial\nu_k}\right)\right](x) = \left[D\left(\frac{\partial}{\partial\nu_j},\frac{\partial}{\partial\nu_k}\right)+D\left(\frac{\partial}{\partial\nu_k},\frac{\partial}{\partial\nu_j}\right)\right](x).$$

According to the condition (RC1) in §4, we have

$$D\left(\frac{\partial}{\partial\nu_j},\frac{\partial}{\partial\nu_k}\right) - D\left(\frac{\partial}{\partial\nu_k},\frac{\partial}{\partial\nu_j}\right) = \left[\frac{\partial}{\partial\nu_j},\frac{\partial}{\partial\nu_k}\right] = 0$$

as in the proof of (4.6). Consequently, we obtain

$$D\left(\frac{\partial}{\partial\nu_j},\frac{\partial}{\partial\nu_k}\right) = D\left(\frac{\partial}{\partial\nu_k},\frac{\partial}{\partial\nu_j}\right)$$

and hence

$$2\left[D\left(\frac{\partial}{\partial\nu_j},\frac{\partial}{\partial\nu_k}\right)\right](x) = \left[D\left(\frac{\partial}{\partial\nu_j}+\frac{\partial}{\partial\nu_k},\frac{\partial}{\partial\nu_j}+\frac{\partial}{\partial\nu_k}\right)\right](x) = 0.$$

This implies $\left[D\left(\frac{\partial}{\partial\nu_j},\frac{\partial}{\partial\nu_k}\right)\right](x) = 0$ and completes the proof of (6.1).▌

Introduce a metric $d : N_x \times N_x \to R$ in N_x by taking

$$d(p,q) = \sqrt{\sum_{i=1}^{n} [\nu_i(p) - \nu_i(q)]^2}$$

for any two points p and q in N_x. On the other hand, for every $p \in N_x$ and every $v \in T(X,p)$, let

$$|v| = \sqrt{\sum_{i=1}^{n} v(\nu_i)}.$$

For every $p \in N_x$ and every positive real number ϵ, the *open ball* $B_\epsilon(p)$ and the *closed ball* $\bar{B}_\epsilon(p)$ are defined by

$$B_\epsilon(p) = \{q \in N_x \mid d(p,q) < \epsilon\}$$
$$\bar{B}_\epsilon(p) = \{q \in N_x \mid d(p,q) \leq \epsilon\}.$$

Since $\nu : N_x \to R^n$ is a homeomorphism of N_x onto the open subspace $\nu(N_x)$ of R^n, $\bar{B}_\epsilon(x)$ is compact when ϵ is sufficiently small.

Lemma 6.2. *For an arbitrary positive real number ϵ such that $\bar{B}_\epsilon(x)$ is compact, there exists a positive real number δ such that, for every point p in $\bar{B}_\epsilon(x)$, the exponential map $\exp_p$ sends an open neighborhood J_p of O_p in $T(X,p)$ diffeomorphically onto the open ball $B_\delta(p)$.*

PROOF. According to (5.9), there exists for every point $p \in \bar{B}_\epsilon(x)$ a positive real number δ_p such that the smooth map

$$h : W \to X \times X$$

constructed in the preceding section is a diffeomorphism on the open neighborhood H_p of O_p in $T(X)$ that consists of all $v \in T(X)$ satisfying

$$d[\pi(v),p] < \delta_p \qquad d[\pi(v), \exp(v)] < \delta_p.$$

The open ball

$$B(p) = B_{\delta_p}(p)$$

is an open neighborhood of p in N_x. Since $\bar{B}_\epsilon(x)$ is compact, there exist a finite number of points $p_1, \ldots, p_m$ in $\bar{B}_\epsilon(x)$ such that

$$\bar{B}_\epsilon(x) = B(p_1) \cup \cdots \cup B(p_m).$$

Let $\delta_i = \delta_{p_i}$ for every $i = 1, 2, \ldots, m$ and

$$\delta = \min \{\delta_1, \ldots, \delta_m\}.$$

Now let $p \in \bar{B}_\epsilon(x)$ be arbitrarily given. Then there exists a positive integer $i \leq m$ with $p \in B(p_i)$. Since h is a diffeomorphism on H_{p_i} and $\delta \leq \delta_i$, it follows that h is a diffeomorphism on the open subspace J of H_{p_i} which consists of all $v \in T(X)$ satisfying

$$d[\pi(v),p_i] < \delta_i \qquad d[\pi(v), \exp\,(v)] < \delta.$$

Let $J_p = J \cap T(X,p)$. Then

$$h(v) = [p, \exp_p(v)]$$

for every $v \in J_p$. Since $h \mid J$ is a diffeomorphism, this implies that $\exp_p$ sends J_p diffeomorphically onto the open ball $B_\delta(p)$.∎

The following corollary is a direct consequence of (6.2).

Corollary 6.3. *For every point x of a Riemannian n-manifold X, there exists an open neighborhood U of x in X such that, for every point $p \in U$, the exponential map* $\exp_p$ *sends an open neighborhood J_p of O_p in $T(X,p)$ diffeomorphically onto U.*

Now let us define a function

$$f : N_x \to$$

by taking

$$f(p) = d(p,x) = \sqrt{\sum_{i=1}^{n} [\nu_i(p)]^2}$$

for every $p \in N_x$. Then f is smooth in $N_x \setminus \{x\}$.

Lemma 6.4. *There exists a positive real number θ such that, for every nontrivial geodesic*

$$\sigma : [-r,r] \to N_x$$

satisfying

$$0 < f[\sigma(0)] \leq \theta \qquad [\sigma_*(0)](f) = 0,$$

the composed function

$$g = f \circ \sigma : [-r,r] \to R$$

has a strict relative minimum $f[\sigma(0)]$ at the point $t = 0$ of $[-r,r]$.

PROOF. On the basis of (6.1), there exists a positive real number $\theta \leq 1$ such that

$$|\Gamma^i_{jk}(p)| < n^{-3}$$

holds for every $i, j, k = 1, 2, \ldots, n$ and every point $p \in N_x$ satisfying

$$d(p,x) = f(p) \leq \theta.$$

Now let $\sigma : [-r,r] \to N_x$ denote any nontrivial geodesic as described in the statement of (6.4). Since g is smooth in an open neighborhood of 0 in $[-r,r]$ and

$$g'(0) = [\sigma_*(0)](f) = 0$$

it suffices to establish $g''(0) > 0$.

Since σ is nontrivial, we have $\sigma_*(0) \neq 0$. We may assume without loss of generality that

$$|\sigma_*(0)| = 1.$$

For each $t \in [-r,r]$, we have the tangent vector

$$\sigma_*(t) = \sum_{i=1}^{n} a_i(t) \frac{\partial}{\partial \nu_i}$$

where a_i denotes the derivative

$$a_i = \frac{d}{dt}(\nu_i \circ \sigma)$$

for every $i = 1, 2, \ldots, n$. This implies that

$$g'(t) = [\sigma_*(t)](f) = \sum_{i=1}^{n} a_i(t) \frac{\partial f}{\partial \nu_i}[\sigma(t)] = \frac{1}{g(t)} \sum_{i=1}^{n} a_i(t)\,\nu_i[\sigma(t)]$$

for every real number t in an open neighborhood U of 0 in $[-r,r]$. In other words,

$$g' = \frac{1}{g} \sum_{i=1}^{n} a_i(\nu_i \circ \sigma)$$

in U. Differentiating this with respect to U, we obtain

$$g'' = \frac{1}{g}\sum_{i=1}^{n}[a_i'(\nu_i \circ \sigma) + a_i^2] - \frac{g'}{g^2}\sum_{i=1}^{n} a_i(\nu_i \circ \sigma).$$

Since $g'(0) = 0$, this implies that

$$g''(0) = \frac{1}{g(0)}\sum_{i=1}^{n}\{a_i'(0)\nu_i[\sigma(0)] + [a_i(0)]^2\}.$$

Since σ is a geodesic, it follows from the preceding section that

$$a_i'(0) = -\sum_{j,k=1}^{n} a_j(0)a_k(0)\Gamma_{jk}^i[\sigma(0)]$$

for every $i = 1, 2, \ldots, n$. On the other hand,

$$\sum_{i=1}^{n}[a_i(0)] = |\sigma_*(0)|^2 = 1.$$

Consequently, we obtain

$$g''(0) = \frac{1}{g(0)}\left\{1 - \sum_{i,j,k=1}^{n} a_j(0)a_k(0)\nu_i[\sigma(0)]\Gamma_{jk}^i[\sigma(0)]\right\}.$$

By the choice of the positive real number θ,

$$|\nu_i[\sigma(0)]| \leq f[\sigma(0)] \leq 1 \qquad |\Gamma_{jk}^i[\sigma(0)]| < n^{-3}$$

for every $i, j, k = 1, 2, \ldots, n$. On the other hand,

$$|a_j(0)| \leq |\sigma_*(0)| = 1$$

for every $j = 1, 2, \ldots, n$. Hence,

$$\left|\sum_{i,j,k=1}^{n} a_j(0)a_k(0)\nu_i[\sigma(0)]\Gamma_{jk}^i[\sigma(0)]\right| \leq \sum_{i,j,k=1}^{n}\left|\Gamma_{jk}^i[\sigma(0)]\right| < 1.$$

Since $g(0) > 0$, this implies that $g''(0) > 0$ and completes the proof of (6.4).▌

Theorem 6.5. *For every point x of a Riemannian n-manifold X, every neighborhood U of x contains a convex open neighborhood V of x. Moreover, V can be so chosen that, for every point p of any convex open subset C of V, the exponential map $\exp_p$ sends a star-shaped open neighborhood W of O_p in $T(X,p)$ onto C diffeomorphically.*

PROOF. Choose a positive real number

$$\epsilon \leq \frac{\theta}{2}$$

such that the closed ball $\bar{B}_\epsilon(x)$ is compact and is contained in U. By (6.2), there exists a positive real number

$$\delta \leq \epsilon$$

such that, for every $p \in \bar{B}_\epsilon(x)$, $\exp_p$ sends an open neighborhood J_p of O_p in $T(X,p)$ diffeomorphically onto $B_\delta(p)$.

Now let us prove that the open ball $B_{\delta/2}(x)$ is convex. For this purpose, let p and q denote any two points in $B_{\delta/2}(x)$. Then

$$d(p,q) \leq d(p,x) + d(q,x) < \frac{\delta}{2} + \frac{\delta}{2} = \delta.$$

Hence $q \in B_\delta(p)$. Since $\exp_p$ sends J_p diffeomorphically onto $B_\delta(p)$, there exists a unique geodesic

$$\sigma : [0,r] \to B_\delta(p)$$

with $\sigma(0) = p$ and $\sigma(r) = q$. It remains to prove that

$$f[\sigma(t)] < \frac{\delta}{2}$$

holds for every $t \in [0,r]$.

For this purpose, let s denote a real number in $[0,r]$ at which the continuous function

$$g = f \circ \sigma : [0,r] \to R$$

attains its maximum value. Then we have

$$g(s) = f[\sigma(s)] = d[x,\sigma(s)] \leq d(x,p) + d[p,\sigma(s)] < \frac{\delta}{2} + \delta < \theta.$$

Assume that $g(s) \geq \delta/2$. Then $s \neq 0$ and $s \neq r$. Hence we must have

$$[\sigma_*(s)](f) = g'(s) = 0.$$

According to (6.4), it follows that g has a strict relative minimum at s. This contradicts our statement that g attains its maximum value at s. Hence we must have $g(s) < \delta/2$. This implies that σ is in $B_{\delta/2}(x)$. This completes the proof of the first part of (6.5) with

$$V = B_{\delta/2}(x).$$

To establish the second part of (6.5), let p denote any point of an arbitrary convex open subset C of V. Then we have

$$C \subset B_\delta(p).$$

It follows that $\exp_p$ sends an open neighborhood S of O_p in $T(X,p)$ onto C diffeomorphically. Since C is convex, it is obvious that S must be star-shaped with O_p as its center. This completes the proof of (6.5).∎

EXERCISES

6A. Show that all results given in the text of this section are true for any smooth n-manifold X with a torsion-free connection D as defined in Exercise 4A.

6B. Prove that every point x of any Riemannian n-manifold X has a neighborhood N_x such that, for every $p \in N_x$, the geodesic from x to p is (except for reparametrization) the only piecewise smooth curve in X from x to p with minimum length. Then prove that a piecewise smooth curve

$$\sigma : [a,b] \rightarrow X$$

is a reparametrization of a geodesic if its length $|\sigma|$ is equal to the distance $\rho[\sigma(a),\sigma(b)]$ between its end points $\sigma(a)$ and $\sigma(b)$.

6C. Consider an arbitrary isometry $f : X \rightarrow Y$ of a Riemannian manifold X onto a Riemannian manifold Y. Prove that the image $f(C)$ of a convex subset C of X is a convex subset of Y.

6D. Prove that a diffeomorphism $f : X \rightarrow X$ of a Riemannian manifold X onto itself is an isometry iff it preserves distance; that is, iff

$$\rho[f(p),f(q)] = \rho(p,q)$$

holds for any two points p and q of X.

6E. Investigate possible generalizations of the concepts and the results in this section to the case where the Riemannian manifold X is infinite-dimensional.

6F. Investigate the notions in this chapter and answer the following two questions:

(a) Which of these notions are *isometric invariants?*

(b) Which of these notions are *diffeomorphism invariants?*

CHAPTER IV

De Rham's Theorem

This final chapter is devoted to the proof as well as the formulation of De Rham's theorem. Singular homology and cohomology groups are defined in the first two sections to the extent that they are needed in the formulation and the proof of De Rham's theorem. The lengthy third section consists of the statement of De Rham's theorem, followed by its proof by means of cohomology with coefficients in presheaves. This proof is based on the one given by Eckmann in [Ec, pp. 72–84]. Other approaches to, and slightly stronger statements of, the theorem are formulated in the exercises.

1. Singular Homology Groups

The singular theory of homology and cohomology is constructed by means of (continuous) maps of some standard n-simplex Δ_n for every non-negative integer n. For convenience of integration over Δ_n, which will be studied in the exercises of this chapter, we deviate from our customary definition, as used in [H1] and define Δ_n as follows: By the *unit n-simplex* of the n-dimensional Euclidean space R^n, we mean the subspace Δ_n of R^n that consists of all points $(x_1, \ldots, x_n)$ of R^n satisfying

$$\sum_{i=1}^{n} x_i \leq 1, \qquad x_i \geq 0 \qquad (i = 1, 2, \ldots, n).$$

As a bounded closed subspace of R^n, Δ_n is a compact metrizable space.

The origin $v_0 = (0, \ldots, 0)$ is a point of Δ_n and will be referred to as the *0th vertex*, or the *leading vertex*, of Δ_n. On the other hand, for each integer $i = 1, 2, \ldots, n$, the point

$$v_i = (\delta_{i1}, \ldots, \delta_{in})$$

of R^n with

$$\delta_{ij} = \begin{cases} 1 & (\text{if } i = j) \\ 0 & (\text{if } i \neq j) \end{cases}$$

is a point of Δ_n and will be referred to as the *ith vertex* of Δ_n. The unit n-simplex Δ_n has $n + 1$ vertices $v_0, v_1, \ldots, v_n$ and is the convex hull of the set

$$V = \{v_0, v_1, \ldots, v_n\}.$$

The closed subspace

$$\Delta_n^{(0)} = \left\{(x_1, \ldots, x_n) \in \Delta_n \,\middle|\, \sum_{i=1}^{n} x_i = 1\right\}$$

of the unit n-simplex Δ_n is called the *0th face* of Δ_n, the *leading face* of Δ_n, or the *face of Δ_n opposite to the vertex v_0*. On the other hand, for each integer $i = 1, 2, \ldots, n$, the closed subspace

$$\Delta_n^{(i)} = \{(x_1, \ldots, x_n) \in \Delta_n \mid x_i = 0\}$$

of Δ_n is called the *ith face* of Δ_n or the *face of Δ_n opposite to the vertex v_i*. Obviously, v_i is not in $\Delta_n^{(i)}$ for every $i = 0, 1, \ldots, n$.

For completeness, we define the *unit 0-simplex* Δ_0 as the singleton space that consists of the real number 0. Δ_n has one vertex $v_0 = 0$ and no face.

Now assume $n > 0$ and consider the maps

$$\kappa_i : \Delta_{n-1} \to \Delta_n \qquad (i = 0, 1, \ldots, n)$$

defined by

$$\kappa_0(x_1, \ldots, x_{n-1}) = \left(1 - \sum_{j=1}^{n-1} x_j, x_1, \ldots, x_{n-1}\right)$$

$$\kappa_i(x_1, \ldots, x_{n-1}) = (x_1, \ldots, x_{i-1}, 0, x_i, \ldots, x_{n-1}) \qquad (i > 0)$$

for every point $(x_1, \ldots, x_{n-1})$ of Δ_{n-1}. One can easily verify that, for every $i = 0, 1, \ldots, n$, κ_i is an imbedding of Δ_{n-1} into Δ_n and its image

$$\kappa_i(\Delta_{n-1}) = \Delta_n^{(i)}$$

is the ith face of Δ_n.

Next assume $n > 1$ and consider the imbeddings

$$\Delta_{n-2} \xrightarrow{\kappa_j} \Delta_{n-1} \xrightarrow{\kappa_i} \Delta_n$$

where $0 \leq i \leq n$ and $0 \leq j \leq n - 1$. The following lemma can be easily verified.

Lemma 1.1. *If $0 \leq j < i \leq n$, then*

$$\kappa_i \circ \kappa_j = \kappa_j \circ \kappa_{i-1} : \Delta_{n-2} \to \Delta_n.$$

Now let X denote an arbitrary topological space and n any nonnegative integer. By a *singular n-simplex* in X, we mean a map

$$\xi : \Delta_n \to X$$

of the unit n-simplex Δ_n into the given topological space X. Here the word "singular" means that ξ does not have to be an imbedding. Let

$$S_n(X) = \text{Map }(\Delta_n, X)$$

denote the set of all singular n-simplexes in X. If $m \neq n$, then by definition we have

$$S_m(X) \cap S_n(X) = \square.$$

The union

$$S(X) = \cup_{n=0}^{\infty} S_n(X)$$

is the set of all singular simplexes in the given space X and will be referred to as the *singular complex* of X.

Assume $n > 0$ and let

$$\xi : \Delta_n \to X$$

denote an arbitrary singular n-simplex in X. For every integer $i = 0, 1, \ldots, n$, the composition

$$\xi \circ \kappa_i : \Delta_{n-1} \to X$$

of the maps

$$\Delta_{n-1} \xrightarrow{\kappa_i} \Delta_n \xrightarrow{\xi} X$$

is a singular $(n-1)$-simplex in X, which will be called the *ith face* of ξ and will be denoted by

$$\xi^{(i)} = \xi \circ \kappa_i.$$

The following proposition is a direct consequence of (1.1).

Proposition 1.2. *If $n > 1$ and $0 \leq j < i \leq n$, then*

$$[\xi^{(i)}]^{(j)} = [\xi^{(j)}]^{(i-1)}$$

for every singular n-simplex ξ in X.

For every nonnegative integer n, let $C_n(X)$ denote the free Abelian group generated by the set $S_n(X)$. Then

$$S_n(X) \subset C_n(X).$$

The elements of the group $C_n(X)$ may be considered as the finite linear forms

$$a_1\xi_1 + a_2\xi_2 + \cdots + a_k\xi_k$$

where $\{\xi_1, \xi_2, \ldots, \xi_k\}$ is a finite subset of $S_n(X)$ and $a_1, a_2, \ldots, a_k$ are integers.

For every $n > 0$, define a function

$$\sigma : S_n(X) \to C_{n-1}(X)$$

by taking

$$\sigma(\xi) = \sum_{i=0}^{n} (-1)^i \xi^{(i)}$$

for every singular n-simplex $\xi : \Delta_n \to X$. Since $C_n(X)$ is the free Abelian group generated by the set $S_n(X)$, it follows that σ extends to a unique homomorphism

$$\partial : C_n(X) \to C_{n-1}(X)$$

which is called the *boundary operator.*

For completeness, we also define

$$C_n(X) = 0$$

for all integers $n < 0$ and define

$$\partial : C_n(X) \to C_{n-1}(X)$$

as the trivial homomorphism whenever $n \leq 0$.

The following proposition can be easily verified by means of (1.2). A detailed proof of it can be found in [H1, p. 211].

Proposition 1.3. *For every integer n, the composed homomorphism*

$$\partial^2 = \partial \circ \partial : C_n(X) \to C_{n-2}(X)$$

of the boundary operators

$$C_n(X) \xrightarrow{\partial} C_{n-1}(X) \xrightarrow{\partial} C_{n-2}(X)$$

is a trivial homomorphism; in symbols,

$$\partial^2 = 0.$$

Thus we obtain a *semi-exact sequence*

$$\cdots \to C_{n+1}(X) \xrightarrow{\partial} C_n(X) \xrightarrow{\partial} C_{n-1}(X) \to \cdots$$

which will be referred to as the *integral singular chain complex* of the space X and will be denoted by the symbol $C(X)$.

The elements of the group $C_n(X)$ are called the *n-dimensional integral singular chains* in the space X. For every $\gamma \in C_n(X)$, the element $\partial(\gamma)$ of $C_{n-1}(X)$ is called the *boundary* of the chain γ.

For every integer n, the kernel of the boundary operator

$$\partial : C_n(X) \to C_{n-1}(X)$$

is called the *group of n-dimensional integral singular cycles of X;* it will be denoted by $Z_n(X)$. On the other hand, the image of the boundary operator

$$\partial : C_{n+1}(X) \to C_n(X)$$

is called the *group of n-dimensional singular boundaries of X;* it will be denoted by $B_n(X)$. Since $C(X)$ is semi-exact, it follows that

$$B_n(X) \subset Z_n(X).$$

The quotient group

$$H_n(X) = Z_n(X)/B_n(X)$$

is called the *n-dimensional integral singular homology group* of the given topological space X.

Now let us establish the following lemma:

Lemma 1.4. *If U is a star-shaped subspace of the n-dimensional Euclidean space R^n with the origin O of R^n as its center, then there exists a homomorphism*

$$h_k : C_k(U) \to C_{k+1}(U)$$

for every integer $k \geq 0$ such that

$$h_{k-1} \circ \partial + \partial \circ h_k = i$$

holds in the following diagram

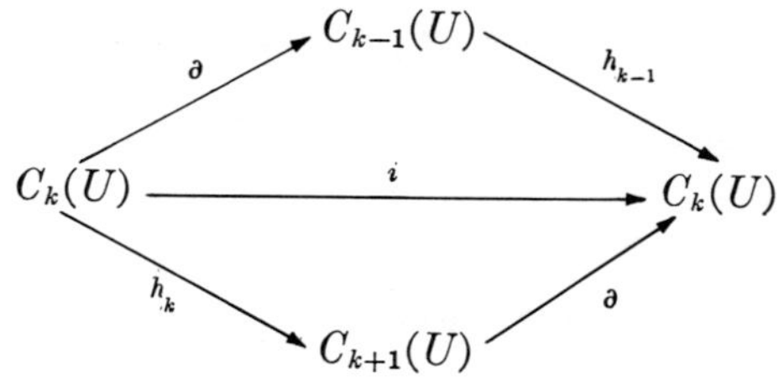

for every $k > 0$, where i stands for the identity homomorphism on $C_k(U)$.

PROOF. Let $k \geq 0$ be arbitrarily given. For every k-dimensional singular simplex $\xi : \Delta_k \to U$, define a $(k + 1)$-dimensional singular simplex $\xi^* : \Delta_{k+1} \to U$ by taking

$$\xi^*(x_1, x_2, \ldots, x_{k+1}) = \Big(\sum_{j=1}^{k+1} x_j\Big)\, \xi(x_2, \ldots, x_{k+1})$$

for every point $(x_1, x_2, \ldots, x_{k+1})$ of the unit $(k + 1)$-simplex Δ_{k+1}. The assignment $\xi \to \xi^*$ defines a function

$$f_k : S_k(U) \to C_{k+1}(U).$$

Since $C_k(U)$ is the free Abelian group generated by the set $S_k(U)$, f_k extends to a unique homomorphism

$$h_k : C_k(U) \to C_{k+1}(U).$$

It is straightforward to verify that

$$h_{k-1} \circ \partial + \partial \circ h_k = i$$

holds for every $k > 0$. This completes the proof of (1.4).∎

The singular $(k + 1)$-simplex ξ^* in the proof of (1.4) is called the *cone obtained by joining ξ to the center O of U.*

The following corollary is a direct consequence of (1.4).

Corollary 1.5. *If U is a star-shaped subspace of the n-dimensional Euclidean space R^n with the origin O of R^n as its center, then we have*

$$H_k(U) = 0$$

for every integer $k > 0$.

In the remainder of this section, we are concerned with an arbitrary topological space X and a given open cover γ of the space X. A singular simplex

$$\xi : \Delta_n \to X$$

is said to be *small* (with respect to γ) iff there exists an open set $U \in \gamma$ with

$$\xi(\Delta_n) \subset U.$$

For every integer $n \geq 0$, the small singular n-simplexes in X generate a subgroup

$$C_n(X;\gamma)$$

of the group $C_n(X)$. Obviously, the boundary operator

$$\partial : C_n(X) \to C_{n-1}(X)$$

sends $C_n(X;\gamma)$ into $C_{n-1}(X;\gamma)$ for every $n > 0$. For completeness, we also define

$$C_n(X;\gamma) = C_n(X) = 0$$

for every integer $n < 0$.

Proposition 1.6. *For every integer n, there exist two homomorphisms*

$$\tau_n : C_n(X) \to C_n(X;\gamma)$$
$$\Omega_n : C_n(X) \to C_{n+1}(X)$$

satisfying the following three conditions:

(a) *For every integer n and every element ξ of $C_n(X)$, we have*

$$\partial[\tau_n(\xi)] = \tau_{n-1}[\partial(\xi)].$$

(b) $\Omega_n(\xi) = 0$ *holds for every integer n and every element ξ of $C_n(X;\gamma)$.*
(c) *For every integer n and every element ξ of $C_n(X)$, we have*

$$\Omega_{n-1}[\partial(\xi)] + \partial[\Omega_n(\xi)] = \tau_n(\xi) - \xi.$$

A proof of (1.6) is contained in the proof of [H2, Chapter II, Lemma 4.3].

EXERCISES

1A. For an arbitrary topological space X, show that $S_0(X)$ can be identified with X itself and that $S_1(X)$ can be identified with the set $P(X)$ of all continuous paths $\xi : I \to X$ in X.

1B. For an arbitrary open cover γ of a topological space X, define the n-dimensional homology group $H_n(X;\gamma)$ of the semi-exact sequence

$$\cdots \to C_{n+1}(X;\gamma) \xrightarrow{\partial} C_n(X;\gamma) \xrightarrow{\partial} C_{n-1}(X;\gamma) \to \cdots.$$

By means of (1.6), prove that

$$H_n(X;\gamma) \approx H_n(X)$$

holds for every integer n.

1C. Let X denote an arbitrary smooth manifold with τ as its smooth structure. By a *smooth n-simplex* in X, we mean a singular n-simplex

$$\xi : \Delta_n \to X$$

that can be extended to a smooth map

$$\xi^* : U \to X$$

over some open subspace U of R^n containing Δ_n. Let $C_n(X;\tau)$ denote the subgroup of $C_n(X)$ generated by all smooth n-simplexes in X. Prove that the boundary operator

$$\partial : C_n(X) \to C_{n-1}(X)$$

sends $C_n(X;\tau)$ into $C_{n-1}(X;\tau)$. Define $H_n(X;\tau)$ for every integer n as in 1B. Study [Ei] and [H3] and then prove that

$$H_n(X;\tau) \approx H_n(X)$$

holds for every integer n. Thus, the singular homology groups of a smooth manifold can be defined by means of the smooth simplexes.

2. Real Singular Cohomology Groups

Let X denote a topological space and consider the integral singular chain complex $C(X)$ defined in the preceding section. For every integer n, let

$$C^n(X;R) = \text{Hom}\,[C_n(X),R]$$

denote the Abelian group of all homomorphisms of $C_n(X)$ into the Abelian group R. The elements of $C^n(X;R)$ are called the *n-dimensional real singular cochains* in X.

For every $n < 0$, we have

$$C^n(X;R) = 0$$

because of $C_n(X) = 0$. For $n \geq 0$, since $C_n(X)$ is the free Abelian group generated by the set $S_n(X)$ of all n-dimensional singular simplexes in X, it follows that the assignment $\phi \to \phi \mid S_n(X)$ defines an isomorphism

$$i : \text{Hom}\,[C_n(X),R] \approx \text{Fun}\,[S_n(X),R]$$

from Hom $[C_n(X),R]$ onto the Abelian group Fun $[S_n(X),R]$ of all real-valued functions on $S_n(X)$. Hence the n-dimensional real singular cochains in X can also be considered as the real-valued functions on $S_n(X)$.

For every integer n, define a homomorphism

$$\delta : C^n(X;R) \to C^{n+1}(X;R)$$

as follows. If $n < 0$, then $C^n(X;G) = 0$ and hence we define $\delta = 0$. Assume $n \geq 0$. Let $\phi \in C^n(X;R)$. Then

$$\phi : C_n(X) \to R$$

is a homomorphism and the composition

$$\delta(\phi) = \phi \circ \partial : C_{n+1}(X) \to R$$

of the boundary operator $\partial : C_{n+1}(X) \to C_n(X)$ is also a homomorphism and hence an element of $C^{n+1}(X;R)$. One can easily verify that $\phi \to \delta(\phi)$ defines a homomorphism. This completes the construction of

$$\delta : C^n(X;R) \to C^{n+1}(X;R)$$

which will be referred to as the *coboundary operator*. For every singular $(n+1)$-simplex ξ and every $\phi \in C^n(X;R)$, we clearly have

$$[\delta(\phi)](\xi) = \sum_{i=0}^{n+1} (-1)^i \phi[\xi^{(i)}].$$

Lemma 2.1. *The composition $\delta \circ \delta$ of the coboundary operators*

$$C^{n-1}(X;R) \xrightarrow{\delta} C^n(X;R) \xrightarrow{\delta} C^{n+1}(X;R)$$

is the trivial homomorphism of $C^{n-1}(X;R)$ into $C^{n+1}(X;R)$; in symbols, $\delta \circ \delta = 0$.

PROOF. If $n < 1$, then $C^{n-1}(X;R) = 0$ and hence we must have $\delta \circ \delta = 0$. Assume $n \geq 1$. Let $\phi \in C^{n-1}(X;R)$ be arbitrarily given. Then we have

$$(\delta \circ \delta)(\phi) = \delta[\delta(\phi)] = \delta(\phi \circ \partial) = \phi \circ \partial \circ \partial = 0$$

because $\partial \circ \partial = 0$. Hence we obtain $\delta \circ \delta = 0$.▮

Thus we obtain a semi-exact sequence

$$\cdots \to C^{n-1}(X;R) \xrightarrow{\delta} C^n(X;R) \xrightarrow{\delta} C^{n+1}(X;R) \to \cdots$$

which will be referred to as the *real singular cochain complex* of the space X and will be denoted by the symbol $C^*(X;R)$.

For every integer n, the kernel of the coboundary operator

$$\delta : C^n(X;R) \to C^{n+1}(X;R)$$

is called the *group of n-dimensional real singular cocycles* in X and will be denoted by $Z^n(X;R)$. On the other hand, the image of the coboundary operator

$$\delta : C^{n-1}(X;R) \to C^n(X;R)$$

is called the *group of n-dimensional real singular coboundaries* in X and will be denoted by $B^n(X;R)$. Since $C^*(X;R)$ is semiexact, it follows that the inclusion

$$B^n(X;R) \subset Z^n(X;R)$$

holds. The quotient group

$$H^n(X;R) = Z^n(X;R)/B^n(X;R)$$

is called the *n-dimensional real singular cohomology group* of the given space X.

Proposition 2.2. *If U is a star-shaped subspace of the n-dimensional Euclidean space R^n with the origin O of R^n as its center, then we have*

$$H^k(U;R) = 0$$

for every integer $k > 0$.

PROOF. Let $k > 0$ be arbitrarily given and consider any k-dimensional real singular cocycle ϕ in U. Then ϕ is a homomorphism

$$\phi : C_k(U) \to R$$

with $\phi \circ \partial = \delta(\phi) = 0$. Using the homomorphism h_{k-1} in (1.4), we obtain a homomorphism

$$\psi = \phi \circ h_{k-1} : C_{k-1}(U) \to R.$$

ψ is an element of $C^{k-1}(U;R)$. By (1.4), we have

$$i = h_{k-1} \circ \partial + \partial \circ h_k.$$

This implies

$$\phi = \phi \circ i = \phi \circ h_{k-1} \circ \partial + \phi \circ \partial \circ h_k = \psi \circ \partial = \delta(\psi)$$

since $\phi \circ h_{k-1} = \psi$ and $\phi \circ \partial = 0$. Therefore,

$$Z^k(U;R) = B^k(U;R).$$

This proves that $H^k(U;R) = 0.$ ▌

Next let us consider an arbitrary (continuous) map

$$f : X \to Y$$

from a topological space X into a topological space Y. For every integer n, f induces a homomorphism

$$f^{\#} : C^n(Y;R) \to C^n(X;R)$$

as follows: If $n < 0$, then $C^n(X;R) = 0$ and hence we may take $f^{\#} = 0$. Assume $n \geq 0$. For each singular n-simplex ξ in X, the composition $f \circ \xi$ is a singular n-simplex in Y. Since $C_n(X)$ is the free Abelian group generated by $S_n(X)$, the assignment $\xi \to f \circ \xi$ determines a homomorphism

$$f_{\#} : C_n(X) \to C_n(Y).$$

Then we define $f^{\#}$ by taking

$$f^{\#}(\phi) = \phi \circ f_{\#} : C_n(X) \to R$$

for every element $\phi : C_n(Y) \to R$ of $C^n(Y;R)$. One can easily verify that $f^{\#}$ sends $Z^n(Y;R)$ into $Z^n(X;R)$ and sends $B^n(Y;R)$ into $B^n(X;R)$. Hence $f^{\#}$ induces a homomorphism

$$f^* = H^n(f;R) : H^n(Y;R) \to H^n(X;R)$$

for every integer n. This homomorphism f^* will be referred to as the *induced homomorphism of the map* $f : X \to Y$ on the n-dimensional real singular cohomology group.

The following two propositions are obvious.

Proposition 2.3. *The induced homomorphism*

$$i^* = H^n(i;R) : H^n(X;R) \to H^n(X;R)$$

of the identity map $i : X \to X$ *on a topological space* X *is the identity automorphism of* $H^n(X;R)$.

Proposition 2.4. *For arbitrary maps*

$$f : X \to Y \qquad g : Y \to Z$$

of topological spaces X, Y, *and* Z, *we have*

$$H^n(g \circ f;R) = H^n(f;R) \circ H^n(g;R).$$

The following corollary of (2.3) and (2.4) can be proved precisely as for the corollary (I,5.8) of (I,5.6) and (I,5.7).

Corollary 2.5. *If* $f : X \to Y$ *is a homeomorphism, then the induced homomorphism*

$$H^n(f;R) : H^n(Y;R) \to H^n(X;R)$$

is an isomorphism.

Now let us consider an arbitrary topological space X together with a given open cover γ of X. Consider the groups $C_n(X;\gamma)$ defined in §1 and let

$$C^n(X;\gamma,R) = \text{Hom}\,[C_n(X;\gamma),R]$$

for every integer n. Then we have

$$C^n(X;\gamma,R) = 0$$

for every $n < 0$. For $n \geq 0$, $C^n(X;\gamma,R)$ can be identified with the Abelian group of all real-valued functions on the set $S_n(X;\gamma)$ of all small singular n-simplexes in X.

Since the boundary operator ∂ sends $C_n(X;\gamma)$ into $C_{n-1}(X;\gamma)$ for every integer n, we may define a homomorphism

$$\delta : C^n(X;\gamma,R) \to C^{n+1}(X;\gamma,R)$$

for every integer n by taking

$$[\delta(\phi)](\xi) = \phi[\partial(\xi)]$$

for every $\phi \in C^n(X;\gamma,R)$ and every $\xi \in C_{n+1}(X;\gamma)$. As in (2.1), one can easily verify that

$$\delta \circ \delta = 0.$$

Thus we obtain a semi-exact upper sequence

$$\cdots \to C^{n-1}(X;\gamma,R) \xrightarrow{\delta} C^n(X;\gamma,R) \xrightarrow{\delta} C^{n+1}(X;\gamma,R) \to \cdots$$

which will be denoted by $C^*(X;\gamma,R)$. For every integer n, the kernel of

$$\delta : C^n(X;\gamma,R) \to C^{n+1}(X;\gamma,R)$$

will be denoted by $Z^n(X;\gamma,R)$ and the image of

$$\delta : C^{n-1}(X;\gamma,R) \to C^n(X;\gamma,R)$$

will be denoted by $B^n(X;\gamma,R)$. Since $C^*(X;\gamma,R)$ is semiexact, it follows that the inclusion

$$B^n(X;\gamma,R) \subset Z^n(X;\gamma,R)$$

holds. The quotient group

$$H^n(X;\gamma,R) = Z^n(X;\gamma,R)/B^n(X;\gamma,R)$$

is called the *n-dimensional real singular cohomology group of X based on the small singular simplexes with respect to the open cover γ of X.*

For every integer n, define a homomorphism

$$j : C^n(X;R) \to C^n(X;\gamma,R)$$

by taking the restriction

$$j(\phi) = \phi \mid C_n(X;\gamma)$$

for every element $\phi : C_n(X) \to R$ in $C^n(X;R)$.

Lemma 2.6. *For every integer n, the commutativity relation $\delta \circ j = j \circ \delta$ holds in the following rectangle:*

$$\begin{array}{ccc} C^n(X;R) & \xrightarrow{\delta} & C^{n+1}(X;R) \\ \downarrow{\scriptstyle j} & & \downarrow{\scriptstyle j} \\ C^n(X;\gamma,R) & \xrightarrow{\delta} & C^{n+1}(X;\gamma,R) \end{array}$$

PROOF. Let $\phi : C_n(X) \to R$ denote any element in $C^n(X;R)$. It suffices to prove that

$$(\delta \circ j)(\phi) = (j \circ \delta)(\phi).$$

For this purpose, let ξ denote any element in $C_{n+1}(X;\gamma)$. Then we have

$$[(\delta \circ j)(\phi)](\xi) = j(\phi)[\partial(\xi)] = \phi[\partial(\xi)]$$

$$[(j \circ \delta)(\phi)](\xi) = [j(\phi \circ \partial)](\xi) = \phi[\partial(\xi)].$$

Since ξ is any element of $C_{n+1}(X;\gamma)$, this completes the proof of (2.6).∎

Because of (2.6), j sends $Z^n(X;R)$ into $Z^n(X;\gamma,R)$ and sends $B^n(X;R)$ into $B^n(X;\gamma,R)$. Hence j induces a homomorphism

$$j^* : H^n(X;R) \to H^n(X;\gamma,R).$$

Proposition 2.7. *This homomorphism j^* is an isomorphism for every integer n.*

PROOF. For every integer n, define a homomorphism

$$k : C^n(X;\gamma,R) \to C^n(X;R)$$

by taking the composition

$$k(\phi) = \phi \circ \tau_n : C_n(X) \to R$$

for every element $\phi : C_n(X;\gamma) \to R$ in $C^n(X;\gamma,R)$. Here τ_n stands for the homomorphism in (1.6). Because of the condition (a) in (1.6), k sends $Z^n(X;\gamma,R)$ into $Z^n(X;R)$ and sends $B^n(X;\gamma,R)$ into $B^n(X;R)$. Hence k induces a homomorphism

$$k^* : H^n(X;\gamma,R) \to H^n(X;R).$$

In view of [H4, p. 22, (6.4)], it remains to prove that $k^* \circ j^*$ and $j^* \circ k^*$ are the identity homomorphisms on $H^n(X;R)$ and $H^n(X;\gamma,R)$ respectively.

To verify that $k^* \circ j^*$ is the identity homomorphism on $H^n(X;R)$, let

$\phi \in Z^n(X;R)$ and $\xi \in C_n(X)$ be arbitrarily given. Define an element $\psi \in C^{n-1}(X;R)$ by taking

$$\psi = \phi \circ \Omega_{n-1} : C_{n-1}(X) \to R$$

where Ω_{n-1} stands for the homomorphism in (1.6). Since

$$(k \circ j)(\phi) = [\phi \mid C_n(X;\gamma)] \circ \tau_n$$
$$\tau_n(\xi) = \xi + \Omega_{n-1}[\partial(\xi)] + \partial[\Omega_n(\xi)]$$

we obtain

$$\begin{aligned}[(k \circ j)(\phi)](\xi) &= \phi[\tau_n(\xi)] \\ &= \phi(\xi) + (\phi \circ \Omega_{n-1} \circ \partial)(\xi) + (\phi \circ \partial \circ \Omega_n)(\xi) \\ &= \phi(\xi) + [\delta(\psi)](\xi)\end{aligned}$$

because $\phi \circ \Omega_{n-1} = \psi$ and $\phi \circ \partial = \delta(\phi) = 0$. The fact that ξ is any element of $C_n(X)$ implies that

$$(k \circ j)(\phi) = \phi + \delta(\psi).$$

Since this holds for every $\phi \in Z^n(X;R)$, $k^* \circ j^*$ is the identity homomorphism on $H^n(X;R)$.

To verify that $j^* \circ k^*$ is the identity homomorphism on $H^n(X;\gamma,R)$, let $\phi \in Z^n(X;\gamma,R)$ and $\xi \in C_n(X;\gamma)$ be arbitrarily given. According to (b) in (1.6), we have

$$\Omega_n(\xi) = 0, \qquad \Omega_{n-1}[\partial(\xi)] = 0.$$

Hence we obtain

$$\begin{aligned}[(j \circ k)(\phi)](\xi) &= \phi[\tau_n(\xi)] \\ &= \phi(\xi) + \phi\{\Omega_{n-1}[\partial(\xi)]\} + (\phi \circ \partial)[\Omega_n(\xi)] \\ &= \phi(\xi).\end{aligned}$$

Since ξ is any element of $C_n(X;\gamma)$, this implies that

$$(j \circ k)(\phi) = \phi.$$

Since this holds for every $\phi \in Z^n(X;\gamma,R)$, $j^* \circ k^*$ is the identity homomorphism on $H^n(X;\gamma,R)$.▌

To conclude the present section, we observe that the Abelian groups $C^n(X;R)$, $Z^n(X;R)$, $B^n(X;R)$, $H^n(X;R)$, and so on, are real vector spaces and that the coboundary operators, induced homomorphisms, and others, are linear transformations. The same is true also in the following exercises.

EXERCISES

2A. Show that every real singular n-cocycle $z : C_n(X) \to R$ in $Z^n(X;R)$ sends $B_n(X)$ into 0 and hence defines a homomorphism

$$z^* : H_n(X) \to R.$$

Verify that $z^* = 0$ if $z \in B^n(X;R)$ and hence z^* depends only on the cohomology class $[z]$ in $H^n(X;R)$. Prove that the assignment $[z] \to z^*$ defines an isomorphism

$$i : H^n(X;R) \approx \text{Hom}\,[H_n(X),R].$$

This is known as the *duality isomorphism.*

2B. Let X denote any smooth manifold with smooth structure τ. Let

$$C^n(X;\tau,R) = \text{Hom}\,[C_n(X;\tau),R]$$

for every integer n, where $C_n(X;\tau)$ stands for a group in Exercise 1C. Define a homomorphism

$$\delta : C^n(X;\tau,R) \to C^{n+1}(X;\tau,R)$$

for every integer n by taking

$$[\delta(\phi)](\xi) = \phi[\partial(\xi)]$$

for every $\phi \in C^n(X;\gamma,R)$ and $\xi \in C_{n+1}(X;\tau)$. Verify that $\delta \circ \delta = 0$. This gives rise to the *n-dimensional smooth cohomology group*

$$H^n(X;\tau,R).$$

Prove that the homomorphism

$$j : C^n(X;R) \to C^n(X;\tau,R)$$

defined by taking the restriction

$$j(\phi) = \phi \mid C_n(X;\tau)$$

for every element $\phi : C_n(X) \to R$ in $C^n(X;R)$ induces an isomorphism

$$j^* : H^n(X;R) \approx H^n(X;\tau,R).$$

2C. Let X denote any smooth manifold with smooth structure τ. Consider an arbitrary smooth k-form ω on X. For any smooth k-simplex ξ in X, choose a smooth extension

$$\xi^* : U \to X$$

of ξ over an open neighborhood U of Δ_k in R^k. Let

$$\omega_\xi = [\mathbf{C}^n(\xi^*)](\omega) = f_\xi \wedge dx_1 \wedge \cdots \wedge dx_k$$

denote the induced smooth form in U, where $f_\xi : U \to R$ is a smooth function according to (II,2.4). Show that the Riemann integral

$$\langle \omega,\xi \rangle = \int_{\Delta_k} f_\xi(x_1, \ldots, x_k)\, dx_1 \cdots dx_k$$

does not depend on the choice of the smooth extension ξ^* of ξ. Verify that the assignment $\xi \to \langle \omega, \xi \rangle$ determines a homomorphism

$$\omega^* : C_k(X;\tau) \to R.$$

By definition, ω^* is an element of the group $C^k(X;\tau,R)$. Prove that the assignment $\omega \to \omega^*$ defines a homomorphism

$$h_k : \mathbf{C}^k(X) \to C^k(X;\tau,R).$$

By means of Stokes' theorem for Riemann integration on Δ_{k+1}, prove that the commutativity relation

$$\delta \circ h_k = h_{k+1} \circ d$$

holds in the following rectangle:

$$\begin{array}{ccc} \mathbf{C}^k(X) & \xrightarrow{h_k} & C^k(X;\tau,R) \\ \downarrow d & & \downarrow \delta \\ \mathbf{C}^{k+1}(X) & \xrightarrow{h_{k+1}} & C^{k+1}(X;\tau,R). \end{array}$$

Consequently, h_k induces a homomorphism

$$h^* : \mathbf{H}^k(X) \to H^k(X;\tau,R)$$

which will be referred to as the *de Rham homomorphism* on $\mathbf{H}^k(X)$.

3. De Rham's Theorem

Let X denote an arbitrary compact smooth n-manifold. The present section is devoted to establish the following *de Rham's theorem:*

Theorem 3.1. *For every integer k, the de Rham cohomology group* $\mathbf{H}^k(X)$ *of* X *is isomorphic to the real singular cohomology group* $H^k(X;R)$ *of* X.

By an *admissible cover* of X, we mean a finite open cover γ of X such that every nonempty intersection

$$U_1 \cap U_2 \cap \cdots \cap U_k$$

of a subcollection $\{U_1, U_2, \ldots, U_k\}$ of γ is diffeomorphic to a star-shaped open subspace of the n-dimensional Euclidean space R^n.

Lemma 3.2. *Every compact smooth n-manifold X has an admissible cover.*

PROOF. Since every compact n-manifold can have only a finite number of components, we may assume without loss of generality that X is connected.

According to (III,2.1), there exists a Riemannian structure

$$\theta : X \to \mathcal{G}(X)$$

in X. This Riemannian structure θ makes X a Riemannian manifold. Since X is connected, it follows from (III,§3) that θ gives rise to a metric

$$\rho : X \times X \to R$$

which defines the topology in X.

According to (III,6.5), every point x of X has a convex open neighborhood V_x in X such that every nonempty convex open subset of V_x is diffeomorphic to a star-shaped open subspace of R^n. Let

$$\alpha = \{V_x \mid x \in X\}.$$

As an open cover of the compact metric space (X,ρ), α has a *Lebesgue number* $r > 0$; that is, every subset of X with diameter less than r is contained in some member of α.

Applying (III,6.5), once more, we obtain, for every point $x \in X$, a convex open neighborhood U_x of x in X with diameter

$$\delta(U_x) < \frac{r}{2}.$$

Since X is compact, the open cover

$$\beta = \{U_x \mid x \in X\}$$

of X has a finite subcollection γ that covers X.

To prove that γ is an admissible cover of X, consider an arbitrary subcollection $\{U_1, U_2, \ldots, U_k\}$ of γ such that

$$C = U_1 \cap U_2 \cap \cdots \cap U_k \neq \Box.$$

It remains to prove that C is diffeomorphic to a star-shaped open subspace of R^n. For this purpose, consider the union

$$W = U_1 \cup U_2 \cup \cdots \cup U_k.$$

Since $C \neq \Box$ and $\delta(U_i) < r/2$ for every $i = 1, 2, \ldots, k$, then $\delta(W) < r$. According to the definition of Lebesgue numbers of α, there exists a point $x \in X$ with

$$W \subset V_x.$$

Let p and q denote any two points in C. Since $C \subset V_x$ and V_x is convex, there exists a unique geodesic σ in V_x joining p to q. Since $C \subset U_i \subset V_x$ and

U_i is convex, it follows that σ is in U_i for every $i = 1, 2, \ldots, k$. This implies that σ is C. Since $C \subset V_x$, σ is the unique geodesic in C joining p to q. Consequently, C is a convex open subset of V_x. By the choice of V_x, this proves that C is diffeomorphic to a star-shaped open subspace of R^n. ▌

Hereafter, let

$$\gamma = \{U_1, U_2, \ldots, U_m\}$$

denote a given admissible cover of X and let K denote the *nerve* of γ; that is, K is the abstract simplicial complex defined as follows. By a *simplex* in K, we mean a nonempty subcollection

$$\sigma = \{U_{i_0}, U_{i_1}, \ldots, U_{i_q}\}$$

of γ satisfying

$$1 \leq i_0 < i_1 < \cdots < i_q \leq m$$

and

$$U_{i_0} \cap U_{i_1} \cap \cdots \cap U_{i_q} \neq \square.$$

The integer q is called the *dimension* of the simplex σ and we shall also say that σ is a *q-simplex* of K. For every q-simplex σ of K with $q > 0$ and every integer $j = 0, 1, \ldots, q$, the *jth face* of σ is the $(q - 1)$-simplex

$$\sigma^{(j)} = \{U_{i_0}, \ldots, \hat{U}_{i_j}, \ldots, U_{i_q}\}$$

of K, where the circumflex over U_{i_j} means that the open set U_{i_j} is deleted from σ.

Following [G, p. 16], we introduce the concept of presheaves on X. By a *presheaf* on X, we mean a function π that assigns to each open set U of X an Abelian group $\pi(U)$ and to each pair (U,V) of open sets of X with $U \subset V$ a homomorphism

$$\pi_{(U,V)} : \pi(V) \to \pi(U)$$

satisfying the following two conditions:

(P1) For every open set U of X, $\pi_{(U,U)}$ is the identity homomorphism on the group $\pi(U)$.

(P2) For any three open sets U, V, and W of X satisfying $U \subset V \subset W$, we have

$$\pi_{(U,V)} \circ \pi_{(V,W)} = \pi_{(U,W)}.$$

Thus a presheaf on X is a *contravariant functor*

$$\pi : \mathcal{C} \to \mathcal{A}$$

from the category $\mathcal{C}$ of all open sets of X and their inclusion maps into the category $\mathcal{A}$ of all Abelian groups and their homomorphisms.

Examples of Presheaves on X:

(a) *The presheaf of smooth k-forms.* Consider the function

$$\pi : \mathcal{C} \to \mathcal{A}$$

defined as follows: For every open set U of X, let

$$\pi(U) = \mathbf{C}^k(U)$$

be the Abelian group of all smooth k-forms on U. For every inclusion map

$$i : U \to V$$

of an open set U of X into an open set $V \supset U$ of X, let

$$\pi_{(U,V)} = \mathbf{C}^k(i) : \mathbf{C}^k(V) \to \mathbf{C}^k(U)$$

be the homomorphism induced by i. Since the conditions (P1) and (P2) are clearly satisfied, it follows that π is a presheaf on X. This presheaf π on X will be referred to as the *presheaf of smooth k-forms* on X and will be denoted by $\mathbf{C}^k$. Similarly, one may define the *presheaf* $\mathbf{Z}^k$ *of closed smooth k-forms* on X and the *presheaf* $\mathbf{B}^k$ *of exact smooth k-forms* on X.

(b) *The presheaf of real singular k-cochains.* Consider the function

$$\pi : \mathcal{C} \to \mathcal{A}$$

defined as follows: For every open set U of X, let

$$\pi(U) = C^k(U;R)$$

be the Abelian group of all real singular k-cochains on U. For every inclusion map

$$i : U \to V$$

of an open set U of X into an open set $V \supset U$ of X, let

$$\pi_{(U,V)} = C^k(i;R) : C^k(V;R) \to C^k(U;R)$$

be the homomorphism induced by i. Since the conditions (P1) and (P2) are clearly satisfied, it follows that π is a presheaf on X. This presheaf π on X will be referred to as the *presheaf of real singular k-cochains* and will be denoted by C^k_R. Similarly, one may define the presheaf Z^k_R of real singular k-cocycles on X and the presheaf B^k_R of real singular k-coboundaries on X.

(c) *The constant presheaves.* For an arbitrary Abelian group G, consider the function

$$\pi : \mathcal{C} \to \mathcal{A}$$

defined as follows: For every open set U of X, let

$$\pi(U) = G.$$

For every pair (U,V) of open sets of X with $U \subset V$, let

$$\pi_{(U,V)} : \pi(V) \to \pi(U)$$

be the identity homomorphism on G. Since the conditions (P1) and (P2) are clearly satisfied, it follows that π is a presheaf on X. This presheaf π on X is called a *constant presheaf* on X and will be denoted by G in case there is no danger of ambiguity.

Next let us define the cohomology groups of the complex K with coefficients in an arbitrarily given presheaf

$$\pi : \mathcal{C} \to \mathcal{A}$$

on X. For every simplex $\sigma = \{U_{i_0}, \ . \ . \ . \ , U_{i_q}\}$ of K, the nonempty open set

$$|\sigma| = U_{i_0} \cap U_{i_1} \cap \cdot \cdot \cdot \cap U_{i_q}$$

of X will be called the *support* of σ. By a *q-cochain of K over π*, we mean a function ϕ that assigns to every q-simplex σ of K an element $\phi(\sigma)$ of the Abelian group $\pi(|\sigma|)$. The set

$$C^q(K;\pi)$$

of all q-cochains of K over π forms an Abelian group with addition defined by

$$(\phi + \psi)(\sigma) = \phi(\sigma) + \psi(\sigma)$$

for any two q-cochains ϕ, ψ of K over π and every q-simplex σ of K. For completeness, we define

$$C^q(K;\pi) = 0$$

when there is no q-simplex in K.

For every q-cochain ϕ of K over π, define a $(q + 1)$-cochain $\delta(\phi)$ of K over π by taking

$$[\delta(\phi)](\sigma) = \sum_{j=0}^{q+1} (-1)^j \pi_{(|\sigma|,|\sigma^{(j)}|)}[\phi(\sigma^{(j)})]$$

for every $(q + 1)$-simplex σ of K. The assignment $\phi \to \delta(\phi)$ defines a homomorphism

$$\delta : C^q(K;\pi) \to C^{q+1}(K;\pi)$$

for every integer q, which will be referred to as the *coboundary operator* on $C^q(K,\pi)$. It is straightforward to verify

$$\delta \circ \delta = 0.$$

Thus we obtain a semi-exact upper sequence

$$\cdot \ \cdot \ \cdot \to C^{q-1}(K;\pi) \xrightarrow{\delta} C^q(K;\pi) \xrightarrow{\delta} C^{q+1}(K;\pi) \to \cdot \ \cdot \ \cdot$$

which will be denoted by $C^*(K;\pi)$. For every integer q, the kernel of

$$\delta : C^q(K;\pi) \to C^{q+1}(K;\pi)$$

will be denoted by $Z^q(K;\pi)$ and the image of

$$\delta : C^{q-1}(K;\pi) \to C^q(K;\pi)$$

will be denoted by $B^q(K;\pi)$. Since $C^*(K;\pi)$ is semiexact, it follows that the inclusion

$$B^q(K;\pi) \subset Z^q(K;\pi)$$

holds. The quotient group

$$H^q(K;\pi) = Z^q(K;\pi)/B^q(K;\pi)$$

is called the *q-dimensional cohomology group of K over* π. If π is the constant presheaf G on X, then it reduces to the *q-dimensional cohomology group* $H^q(K;G)$ *of the simplicial complex K with coefficients in the Abelian group* as usually defined in combinatorial topology.

Lemma 3.3. *If* π *is the presheaf* $\mathbf{C}^k$ *of smooth k-forms on X, then we have*

$$H^q(K;\pi) = 0$$

or every integer $q > 0$.

PROOF. Let $q > 0$ be arbitrarily given. It suffices to prove

$$Z^q(K;\pi) = B^q(K;\pi).$$

For this purpose, let $\phi \in Z^q(K;\pi)$ be arbitrarily given. Then we have

$$\begin{aligned}\{[\delta(\phi)](\sigma)\}(x) &= \sum_{j=0}^{q+1} (-1)^j \pi_{(|\sigma|,|\sigma^{(j)}|)}[\phi(\sigma^{(j)})](x)\\ &= \sum_{j=0}^{q+1} (-1)^j [\phi(\sigma^{(j)})](x) = 0\end{aligned}$$

for every $(q+1)$-simplex σ of K and every point x in $|\sigma|$.

According to (I,3.11), there exists a smooth partition of unity subordinate to the given admissible cover γ of X. In other words, for every integer $i = 1, 2, \ldots, m$, there is a smooth function

$$f_i : X \to R$$

with support $S(f_i) \subset U_i$ such that

$$\sum_{i=1}^{m} f_i(x) = 1$$

holds for every point x of X. For every integer $i = 1, 2, \ldots, m$, define $\phi_i \in C^q(K;\pi)$ by taking

$$\phi_i(\sigma) = (f_i \mid |\sigma|) \wedge \phi(\sigma) \in \mathbf{C}^k(|\sigma|)$$

for every q-simplex σ of K. Then we have

$$\phi = \sum_{i=1}^{m} \phi_i$$

and

$$\begin{aligned}\{[\delta(\phi_i)](\sigma)\}(x) &= \sum_{j=0}^{q+1} (-1)^j[\phi_i(\sigma^{(j)})](x) \\ &= \sum_{j=0}^{q+1} (-1)^j f_i(x)[\phi(\sigma^{(j)})](x) \\ &= f_i(x)\{[\delta(\phi)](\sigma)\}(x) = 0\end{aligned}$$

for every $(q + 1)$-simplex σ of K and every point x in $|\sigma|$. The last implies that

$$\delta(\phi_i) = 0$$

for every $i = 1, 2, \ldots, m$. Consequently, in order to prove

$$\phi \in B^q(K;\pi)$$

it suffices to construct a $(q - 1)$-cochain

$$\psi_i \in C^{q-1}(K;\pi)$$

for every $i = 1, 2, \ldots, m$ such that

$$\phi_i = \delta(\psi_i).$$

For this purpose, let us first consider the case $i = 1$. We define a $(q - 1)$-cochain

$$\psi_1 \in C^{q-1}(K;\pi)$$

as follows: Let

$$\sigma = \{U_{i_1}, \ldots, U_{i_q}\}$$

denote an arbitrary $(q - 1)$-simplex of K. If $i_1 = 1$, we take $\psi_1(\sigma) = 0$. If $i_1 > 1$ and $U_1 \cap |\sigma| = \square$, we also take $\psi_1(\sigma) = 0$. If $i_1 > 1$ and $U_1 \cap |\sigma| \neq \square$, then

$$\tau = \{U_1, U_{i_1}, \ldots, U_{i_q}\}$$

is a q-simplex of K with $\tau^{(0)} = \sigma$ and

$$\tau^{(j)} = \{U_1, U_{i_1}, \ldots, \hat{U}_{i_j}, \ldots, U_{i_q}\}$$

for every $j = 1, 2, \ldots, q$. In this case, we define

$$\psi_1(\sigma) \in \mathbf{C}^k(|\sigma|)$$

by taking

$$[\psi_1(\sigma)](x) = \begin{cases} [\phi_1(\tau)](x) & (x \in |\tau|) \\ 0 & (x \in |\sigma| \setminus |\tau|). \end{cases}$$

This completes the construction of ψ_1. To prove $\phi_1 = \delta(\psi_1)$, let

$$\tau = \{U_{i_0}, U_{i_1}, \ldots, U_{i_q}\}$$

denote any q-simplex of K and let x denote any point of $|\tau|$. If $i_0 = 1$, we have

$$\begin{aligned} \{[\delta(\psi_1)](\tau)\}(x) &= \sum_{j=0}^{q} (-1)^j[\psi_1(\tau^{(j)})](x) \\ &= [\psi_1(\tau^{(0)})](x) \\ &= [\phi_1(\tau)](x). \end{aligned}$$

If $i_0 > 1$ and $x \notin U_1$, we have

$$\{[\delta(\psi_1)](\tau)\}(x) = 0 = [\phi_1(\tau)](x).$$

If $i_0 > 1$ and $x \in U_1$, then

$$\theta = \{U_1, U_{i_0}, U_{i_1}, \ldots, U_{i_q}\}$$

is a $(q + 1)$-simplex of K. In this case, we have

$$\begin{aligned} \{[\delta(\psi_1)](\tau)\}(x) &= \sum_{j=0}^{q} (-1)^j[\phi_1(U_1, U_{i_0}, \ldots, \hat{U}_{i_j}, \ldots, U_{i_q})](x) \\ &= [\phi_1(\tau)](x) - \{[\delta(\phi_1)](\theta)\}(x) \\ &= [\phi_1(\tau)](x) \end{aligned}$$

since $\delta(\phi_1) = 0$. This proves $\delta(\psi_1) = \phi_1$.

From the definition of K, one can easily see that the order

$$\{U_1, U_2, \ldots, U_m\}$$

of the open sets in the cover γ is inessential and is used only for our convenience. Hence there exists a $(q - 1)$-cochain $\psi_i \in C^{q-1}(K;\pi)$ for every $i = 1, 2, \ldots, m$ with

$$\phi_i = \delta(\psi_i).$$

This completes the proof of (3.3).∎

Lemma 3.4. *If π is the presheaf C^k_R of real singular k-cochains on X, then we have*

$$H^q(K;\pi) = 0$$

for every integer $q > 0$.

PROOF. Let $q > 0$ be arbitrarily given. It suffices to prove that

$$Z^q(K;\pi) = B^q(K;\pi).$$

For this purpose, let $\phi \in Z^q(K;\pi)$ be arbitrarily given. Then we have

$$\{[\delta(\phi)](\sigma)\}(\xi) = \sum_{j=0}^{q+1} (-1)^j \pi_{(|\sigma|,|\sigma^{(j)}|)}[\phi(\sigma^{(j)})](\xi)$$
$$= \sum_{j=0}^{q+1} (-1)^j[\phi(\sigma^{(j)})](\xi) = 0$$

for every $(q+1)$-simplex σ of K and singular k-simplex $\xi : \Delta_k \to X$ with

$$\xi(\Delta_k) \subset |\sigma|.$$

To prove $\phi \in B^q(K;\pi)$, it suffices to construct a $(q-1)$-cochain

$$\psi \in C^{q-1}(K;\pi)$$

with

$$\phi = \delta(\psi).$$

For this purpose, let

$$\sigma = \{U_{i_1}, \ldots, U_{i_q}\}$$

denote an arbitrary $(q-1)$-simplex of K and

$$\xi : \Delta_k \to |\sigma| \subset X$$

denote any singular k-simplex in $|\sigma|$. Let i_0 denote the least integer satisfying

$$\xi(\Delta_k) \subset U_{i_0}.$$

Then we have $i_0 \leq i_1$. If $i_0 = i_1$, we define

$$[\psi(\sigma)](\xi) = 0.$$

If $i_0 < i_1$, then

$$\tau = \{U_{i_0}, U_{i_1}, \ldots, U_{i_q}\}$$

is a q-simplex of K with $\xi(\Delta_k) \subset |\tau|$. In this case, we define

$$[\psi(\sigma)](\xi) = [\phi(\tau)](\xi).$$

This completes the construction of the $(q-1)$-cochain ψ.

To prove that $\phi = \delta(\psi)$, let

$$\sigma = \{U_{i_0}, U_{i_1}, \ldots, U_{i_q}\}$$

denote an arbitrary q-simplex of K and

$$\xi : \Delta_k \to |\sigma| \subset X$$

denote any singular k-simplex in $|\sigma|$. Let i_* denote the least integer satisfying

$$\xi(\Delta_k) \subset U_{i_*}.$$

Then we have $i_* \leq i_0$. If $i_* = i_0$, then

$$[\psi(\sigma^{(0)})](\xi) = [\phi(\sigma)](\xi)$$

and

$$[\psi(\sigma^{(j)})](\xi) = 0 \qquad (j = 1, 2, \ldots, q).$$

In this case, we obtain

$$\{[\delta(\psi)](\sigma)\}(\xi) = \sum_{j=0}^{q} (-1)^j[\psi(\sigma^{(j)})](\xi) = [\phi(\sigma)](\xi).$$

If $i_* < i_0$, then

$$\tau = \{U_{i_*}, U_{i_0}, U_{i_1}, \ldots, U_{i_q}\}$$

is a $(q + 1)$-simplex of K. In this case,

$$\begin{aligned}
\{[\delta(\psi)](\sigma)\}(\xi) &= \sum_{j=0}^{q} (-1)^j[\psi(\sigma^{(j)})](\xi) \\
&= \sum_{j=0}^{q} (-1)^j[\phi(U_{i_*}, U_{i_0}, \ldots, \hat{U}_{i_j}, \ldots, U_{i_q})](\xi) \\
&= -\sum_{j=1}^{q+1} (-1)^j[\phi(\tau^{(j)})](\xi) \\
&= [\phi(\sigma)](\xi) - \{[\delta(\phi)](\tau)\}(\xi) \\
&= [\phi(\sigma)](\xi)
\end{aligned}$$

since $\delta(\phi) = 0$. This completes the proof of $\phi = \delta(\psi)$ and (3.4).∎

Next let us consider any two presheaves

$$\omega, \pi : \mathfrak{C} \to \mathfrak{A}$$

on X. By a *homomorphism*

$$h : \omega \to \pi$$

from ω to π, we mean a homomorphism

$$h_U : \omega(U) \to \pi(U)$$

for every open set U of X such that the commutativity relation

$$h_U \circ \omega_{(U,V)} = \pi_{(U,V)} \circ h_V$$

holds for every pair (U,V) of open sets of X with $U \subset V$.

An arbitrary homomorphism $h : \omega \to \pi$ induces a homomorphism

$$h^* = H^q(K;h) : H^q(K;\omega) \to H^q(K;\pi)$$

for every integer q as follows: First, h induces a homomorphism

$$h^{\#} = C^q(K;h) : C^q(K;\omega) \to C^q(K;\pi)$$

defined for every integer q by

$$[h^{\#}(\phi)](\sigma) = h[\phi(\sigma)]$$

for every q-cochain $\phi \in C^q(K;\omega)$ and every q-simplex σ in K. One can easily verify that $h^{\#}$ commutes with the coboundary operator δ; that is,

$$\delta \circ h^{\#} = h^{\#} \circ \delta.$$

Thus $h^{\#}$ sends $Z^q(K;\omega)$ into $Z^q(K;\pi)$ and $B^q(K;\omega)$ into $B^q(K;\pi)$. Consequently, $h^{\#}$ induces a homomorphism

$$h^* : H^q(K;\omega) \to H^q(K;\pi)$$

for every integer q.

By a *short exact sequence* of presheaves on X, we mean any two homomorphisms

$$\omega \xrightarrow{f} \pi \xrightarrow{g} \theta$$

of presheaves on X such that, for every open set U of X,

$$0 \to \omega(U) \xrightarrow{f_U} \pi(U) \xrightarrow{g_U} \theta(U) \to 0$$

is a short exact sequence of Abelian groups in the sense that f_U is a monomorphism, g_U is an epimorphism, and $\text{Im}\ (f_U) = \text{Ker}\ (g_U)$. For such a short exact sequence (f,g), one can easily verify that

$$0 \to C^q(K;\omega) \xrightarrow{f^{\#}} C^q(K;\pi) \xrightarrow{g^{\#}} C^q(K;\theta) \to 0$$

is a short exact sequence for every integer q. These short exact sequences give rise to a *Bockstein homomorphism*

$$\beta : H^q(K;\theta) \to H^{q+1}(K;\omega)$$

for every integer q as follows: Let ξ be an arbitrary element of $H^q(K;\theta)$ and choose any cocycle

$$z \in C^q(K;\theta)$$

that represents ξ. Since $g^{\#}$ is an epimorphism, there exists a cochain

$$y \in C^q(K;\pi)$$

with $g^{\#}(y) = z$. Consider $\delta(y) \in C^{q+1}(K;\pi)$. Since

$$g^{\#}[\delta(y)] = \delta[g^{\#}(y)] = \delta(z) = 0$$

we have

$$\delta(y) \in \text{Ker}\ (g^{\#}) = \text{Im}\ (f^{\#}).$$

Hence there exists a cochain

$$x \in C^{q+1}(K;\omega)$$

with $f^{\#}(x) = \delta(y)$. Since

$$f^{\#}[\delta(x)] = \delta[f^{\#}(x)] = \delta[\delta(y)] = 0$$

and $f^{\#}$ is a monomorphism, we have $\delta(x) = 0$ and hence

$$x \in Z^{q+1}(K;\omega).$$

One can easily verify that the element $\beta(\xi)$ of $H^{q+1}(K;\omega)$ represented by the cochain x does not depend on the choice of z, y, and x. The assignment $\xi \to \beta(\xi)$ defines the *Bockstein homomorphism*

$$\beta : H^q(K;\theta) \to H^{q+1}(K;\omega)$$

of the given short exact sequence (f,g) for every integer q. Together with the induced homomorphisms

$$f^* : H^q(K;\omega) \to H^q(K;\pi)$$
$$g^* : H^q(K;\pi) \to H^q(K;\theta)$$

for all integers q, these Bockstein homomorphisms form a sequence

$$\cdots \to H^q(K;\omega) \xrightarrow{f^*} H^q(K;\pi) \xrightarrow{g^*} H^q(K;\theta) \xrightarrow{\beta} H^{q+1}(K;\omega) \to \cdots$$

of Abelian groups and homomorphisms, which will be referred to as the *Bockstein cohomology sequence* of the complex K for the given short exact sequence (f,g) of presheaves on X. The following lemma can be established by standard methods used to prove similar theorems in homological algebra, [H4, Chapter I, (6.9)].

Lemma 3.5. *The Bockstein cohomology sequence of the complex K for any short exact sequence (f,g) of presheaves on X is exact.*

The following corollary of (3.5) is a direct consequence of the exactness.

Corollary 3.6. *If, for some integer q, we have*

$$H^q(K;\pi) = 0 \qquad H^{q+1}(K;\pi) = 0$$

then the Bockstein homomorphism

$$\beta : H^q(K;\theta) \to H^{q+1}(K;\omega)$$

is an isomorphism.

As our first example of short exact sequence of presheaves on X, we have the homomorphisms

$$\mathbf{Z}^k \xrightarrow{i} \mathbf{C}^k \xrightarrow{d} \mathbf{B}^{k+1}$$

of the presheaves in Example (a). Here, for every open set U of X, the homomorphisms i_U and d_U in the short exact sequence

$$0 \to \mathbf{Z}^k(U) \xrightarrow{i_U} \mathbf{C}^k(U) \xrightarrow{d_U} \mathbf{B}^{k+1}(U) \to 0$$

of Abelian groups are the inclusion homomorphism and the exterior differentiation respectively. The exactness is obvious.

The following corollary is a direct consequence of (3.3) and (3.6).

Corollary 3.7. *For arbitrary integers $k \geq 0$ and $q > 0$, we have an isomorphism*

$$\beta : H^q(K;\mathbf{B}^{k+1}) \approx H^{q+1}(K;\mathbf{Z}^k).$$

Next let us establish the following lemma:

Lemma 3.8. *For every $k \geq 0$ and every $q \geq 0$, we have*

$$H^q(K;\mathbf{B}^{k+1}) = H^q(K;\mathbf{Z}^{k+1}).$$

PROOF. Let σ denote any q-simplex of K. Since the open cover γ is admissible, it follows from the definition of admissible covers that the support $|\sigma|$ of σ is diffeomorphic to a star-shaped open subspace of the Euclidean n-space R^n. By (II,5.8) and Poincaré's lemma (II,6.1), we have

$$\mathbf{H}^{k+1}(|\sigma|) = 0.$$

This implies that

$$\mathbf{B}^{k+1}(|\sigma|) = \mathbf{Z}^{k+1}(|\sigma|).$$

Since this is true for every q-simplex σ of K, we have

$$C^q(K;\mathbf{B}^{k+1}) = C^q(K;\mathbf{Z}^{k+1})$$

and, consequently,

$$H^q(K;\mathbf{B}^{k+1}) = H^q(K;\mathbf{Z}^{k+1}).$$

This completes the proof of (3.8).▌

The following corollary is a direct consequence of (3.7) and (3.8).

Corollary 3.9. *For every integer $k > 0$, we have isomorphisms*

$$H^1(K;\mathbf{Z}^{k-1}) \approx H^2(K;\mathbf{Z}^{k-2}) \approx \cdots \approx H^k(K;\mathbf{Z}^0).$$

Lemma 3.10. *For every integer $k > 0$, we have the isomorphism*

$$H^1(K;\mathbf{Z}^{k-1}) \approx \mathbf{H}^k(X).$$

PROOF. Since $H^1(K;\mathbf{C}^{k-1}) = 0$, we have the following exact sequence

$$H^0(K;\mathbf{C}^{k-1}) \xrightarrow{d^*} H^0(K;\mathbf{B}^k) \xrightarrow{\beta} H^1(K;\mathbf{Z}^{k-1}) \to 0$$

as a part of the Bockstein cohomology sequence of K for the short exact sequence

$$\mathbf{Z}^{k-1} \xrightarrow{i} \mathbf{C}^{k-1} \xrightarrow{d} \mathbf{B}^k.$$

From the exactness, it follows immediately that β is an epimorphism. By (3.8) and the definition of 0-dimensional cohomology groups of K, one can easily verify the following identifications:

$$H^0(K;\mathbf{B}^k) = H^0(K;\mathbf{Z}^k) = Z^0(K;\mathbf{Z}^k) = \mathbf{Z}^k(X)$$
$$H^0(K;\mathbf{C}^{k-1}) = Z^0(K;\mathbf{C}^{k-1}) = \mathbf{C}^{k-1}(X).$$

After these identifications, the homomorphism d^* becomes the exterior differentiation

$$d : \mathbf{C}^{k-1}(X) \to \mathbf{Z}^k(X).$$

Hence the image of d^* may be identified with the subgroup $\mathbf{B}^k(X)$ of $\mathbf{Z}^k(X)$. Since the kernel of β coincides with the image of d^*, the epimorphism β induces an isomorphism

$$\beta^* : \mathbf{H}^k(X) \approx H^1(K;\mathbf{Z}^{k-1}).$$

This completes the proof of (3.10).▌

Lemma 3.11. *For every integer $k > 0$, we have the isomorphism*

$$H^k(K;\mathbf{Z}^0) \approx H^k(K;R).$$

PROOF. According to the definition of admissible covers of X, the support $|\sigma|$ of any simplex σ in K is connected and hence

$$\mathbf{Z}^0(U) \approx R.$$

This implies (3.11).▌

The following corollary is a direct consequence of (3.9), (3.10), and (3.11).

Corollary 3.12. *For every integer $k > 0$, we have the isomorphism*

$$\mathbf{H}^k(X) \approx H^k(K;R).$$

As our second example of short exact sequence of presheaves on X, we have the homomorphisms

$$Z_R^k \xrightarrow{i} C_R^k \xrightarrow{\delta} B_R^{k+1}$$

of the presheaves in Example (b). Here, for every open set U of X, the homomorphisms i_U and δ_U in the short exact sequence

$$0 \to Z^k(U;R) \xrightarrow{i_U} C^k(U;R) \xrightarrow{\delta_U} B^{k+1}(U;R) \to 0$$

of Abelian groups are the inclusion homomorphism and the coboundary operator respectively. The exactness is obvious.

The following corollary is a direct consequence of (3.4) and (3.6).

Corollary 3.13. *For arbitrary integers $k \geq 0$ and $q > 0$, we have an isomorphism*

$$\beta : H^q(K;B_R^{k+1}) \approx H^{q+1}(K;Z_R^k).$$

The following lemma can be established by the same method as used in the proof of (3.8) with (II,5.8) and (II,6.1) replaced by (2.5) and (2.2) respectively.

Lemma 3.14. *For every $k \geq 0$ and every $q \geq 0$, we have*

$$H^q(K;B_R^{k+1}) = H^q(K;Z_R^{k+1}).$$

The following corollary is a direct consequence of (3.13) and (3.14).

Corollary 3.15. *For every integer $k > 0$, we have isomorphisms*

$$H^1(K;Z_R^{k-1}) \approx H^2(K;Z_R^{k-2}) \approx \cdots \approx H^k(K;Z_R^0).$$

The following lemma can be established by the same method as used in the proof of (3.10).

Lemma 3.16. *For every integer $k > 0$, we have the isomorphism*

$$H^1(K;Z_R^{k-1}) \approx H^k(X;\gamma,R).$$

The following lemma can be established by the same method as used in the proof of (3.11).

Lemma 3.17. *For every integer $k > 0$, we have the isomorphism*

$$H^k(K;Z_R^0) \approx H^k(K;R).$$

The following corollary is a direct consequence of (3.15), (3.16), and (3.17).

Corollary 3.18. *For every integer $k > 0$, we have the isomorphism*

$$H^k(X;\gamma,R) \approx H^k(K;R).$$

Finally, let us establish de Rham's theorem.

PROOF OF (3.1). The theorem is obvious when $k \leq 0$. In fact, we have

$$\mathbf{H}^k(X) = 0 \qquad H^k(X;R) = 0$$

for every $k < 0$. For the case $k = 0$, let m denote the number of components of the compact n-manifold. Then we have

$$\mathbf{H}^0(X) \approx R^m \qquad H^0(X;R) \approx R^m.$$

Assume that $k > 0$. By (3.12) and (3.18),

$$\mathbf{H}^k(X) \approx H^k(X;\gamma,R).$$

On the other hand,

$$H^k(X;R) \approx H^k(X;\gamma,R)$$

according to (2.7). Consequently,

$$\mathbf{H}^k(X) \approx H^k(X;R).$$

This completes the proof.∎

EXERCISES

3A. For every integer k, prove that

$$\mathbf{H}^k(X) \approx \text{Hom}\,[H_k(X),R].$$

3B. By a *smooth triangulation* of X, we mean a homeomorphism

$$h : K \to X$$

from a simplicial complex K onto X such that, for every closed simplex σ of K, $h|\sigma$ is smooth. Find a proof from the literature, say [W, pp. 124–135], for *Cairns' Theorem*, which establishes the existence of a finite smooth triangulation of X.

3C. Let $h : K \to X$ denote any finite smooth triangulation of X. As in Exercise 2C, establish a homomorphism

$$h^* : \mathbf{H}^q(X) \to H^q(K;R)$$

from the de Rham cohomology group $\mathbf{H}^q(X)$ into the real simplicial cohomology group $H^q(K;R)$ of K by means of Riemann integration and Stokes' theorem. Find a proof from the literature, say [S-T, pp. 145–156], for the fact that h^* is an isomorphism for every integer q.

3D. Prove that the de Rham homomorphism

$$h^* : \mathbf{H}^k(X) \to H^k(X;\tau,R)$$

in Exercise 2C is an isomorphism for every integer k. Then deduce de Rham's theorem (3.1) from this result and Exercise 2B.

Bibliography

[A-M] Auslander, L., and R. E. MacKenzie, *Introduction to Differentiable Manifolds.* New York: McGraw-Hill Book Company, Inc., 1963.

[B-C] Bishop, R., and R. J. Crittenden, *Geometry of Manifolds.* New York: Academic Press, Inc., 1964.

[C] Chevalley, C., *The Theory of Lie Groups.* Princeton, N.J.: Princeton University Press, 1946.

[Co] Coddington, E. A., *An Introduction to Ordinary Differential Equations.* Englewood Cliffs, N.J.: Prentice-Hall, Inc., 1961.

[D] De Rham, G., *Varietés Différentiables,* Actualités scientifiques et industrielles 1222. Paris: Hermann & Cie, 1960.

[Ec] Eckmann, B., "Differentiable Manifolds," (notes from a course), University of Michigan, Ann Arbor, Mich., 1960.

[Ei] Eilenberg, S., "*Singular Homology in Differentiable Manifolds,*" *Annals of Mathematics,* vol. 48, pp. 670–681, 1947.

[G] Godement, R., *Théorie des Faisceaux,* Actualités scientifiques et industrielles 1252. Paris: Hermann & Cie, 1958.

[H1] Hu, S.-T., *Homology Theory, A First Course to Algebraic Topology.* San Francisco: Holden-Day, Inc., 1966.

[H2] Hu, S.-T., *Cohomology Theory.* Chicago: Markham Publishing Company, 1968.

[H3] Hu, S.-T., "*On Singular Homology in Differentiable Spaces,*" *Annals of Mathematics,* vol. 50, pp. 266–269, 1949.

[H4] Hu, S.-T., *Introduction to Homological Algebra.* San Francisco: Holden-Day, Inc., 1968.

[L] Lang, S., *Introduction to Differentiable Manifolds.* New York: Interscience Publishers, 1962.

[Mi] Milnor, J., "*On Manifolds Homeomorphic to the 7-Sphere,*" *Annals of Mathematics*, vol. 64, pp. 399–405, 1956.

[Mu] Munkres, J. R., *Elementary Differential Topology*, Annals of Mathematics Studies, #54. Princeton, N.J.: Princeton University Press, 1961.

[S-T] Singer, I. M., and J. A. Thorpe, *Lecture Notes on Elementary Topology and Geometry*. Glenview, Ill.: Scott, Foresman and Company, 1967.

[W] Whitney, H., *Geometric Integration Theory*. Princeton, N.J.: Princeton University Press, 1957.

Index

Index

U

V

W

Z